Practical Biology

A guide to teacher assessment

M K Sands BSc, PhD
and
P E Bishop BSc

Bell & Hyman · London

First published 1984 by
Bell & Hyman Limited
Denmark House
37-39 Queen Elizabeth Street
London SE1 2QB

Sands, M.K.
Practical biology. — (Modern teachers series)
1. Biology — Study and teaching (Higher)
2. Grading and marking (Students)
I. Title II. Bishop, P.E. III. Series
574'.076 QH316

ISBN-0-7135-1405-1

Printed in Great Britain by The Thetford Press, Thetford, Norfolk.

Contents

Acknowledgements

We are grateful to R.J. Whittaker and Edward Arnold (publishers) for permission to reproduce the material given in Appendix 2.

We are also grateful to the following examining boards who have given permission to reproduce questions and appendices as indicated in the text: The Associated Examining Board (AEB), University of Cambridge Local Examinations Syndicate (C), Joint Matriculation Board (JMB), University of London School Examinations Council (L), Oxford Local Examinations (O), Oxford and Cambridge Schools Examination Board (OC), Welsh Joint Education Committee (W).

Investigations 4.2, 6.3 and 6.4 are based on Experiment 6 in *Food Tests*, Experiment 14 in *Diffusion and Osmosis* and Experiment 11 in *Enzymes* by D.G. Mackean (John Murray, 1971 and 1972).

Investigations 7.3 and 7.4 are based on Experiments 10.2 and 4.4 in *Biology, A Functional Approach: Students' Manual* by M.B.V. Roberts (Nelson, 1974).

Figs 6.1 and 6.5 are reprinted from *Advanced Biology* by J. Simpkins and J.I. Williams (Bell and Hyman, 1983).

The authors wish to record thanks to the following: P.W. Freeland, G.R. Millington and C.F. Stoneman for the original ideas on which three investigations are based; J.D. Wark for his photographic work on microslides from Philip Harris Biological Ltd; L. Cummings and K.E. Selkirk for reading the statistics sections in Chapter 8; A. Mower for typing; the class of '83 for the cover photograph by J. Eyett.

The authors also wish to thank the following teachers and their sixth-form students who pre-tested exercises for Chapter 9: D. Collingwood, The Abbot Beyne School, Burton-on-Trent; P. Dawson, Nottingham High School, Nottingham; M. Evans, Queen Elizabeth Girls' School, Mansfield; J.S. Gorman, Sherwood Hall Upper School, Mansfield; E.M. Jones, Nottingham High School for Girls, Nottingham.

Chapter 1

INTERNAL ASSESSMENT OF PRACTICAL SKILLS

In recent years a number of Advanced level examinations in biology have included some form of internal teacher assessment of students' practical abilities. In some cases there is no practical examination at all and the teacher assessment counts for the whole of the student's practical mark. In others, it counts for only part of the assessment and a practical examination is retained.

The introduction of teacher assessment into advanced biology courses causes a variety of reactions from teachers. A few prefer to stick to the tried and trusted methods of the type of practical work which they have long used. More take up the challenge whole-heartedly, recognising that teacher assessment at this level is a logical extension of the educational theories which resulted in such assessment at CSE level, and which were reported as being favoured by science teachers and students in the report, *Practical Work in School Science* by J.F. Kerr et al, as long ago as 1964. Others falter through lack of finance, laboratory equipment or technical help. Still others, while keen to meet the challenge, feel the weight of responsibility press upon them or judge themselves to be insufficiently experienced to cope with the demands, and to teachers new to the classroom or to advanced work internal assessment must seem daunting.

Most teachers seem to be agreed on the reasons for and aims of including practical work in their courses. Research enquiries into the views and practices of teachers with regard to practical work in school science have underlined this agreement. A brief look at some of the reasons will enable the reader to consider the extent to which teacher assessment of ongoing practical work helps in the achievement of these aims more than does an externally set practical examination at the end of the course.

Practical work is seen as providing a way not only of developing a number of different important practical skills such as those detailed in this book, but also favourable attitudes: it provides interest and enjoyment, produces enthusiasm, encourages initiative, imagination and cooperation, and develops self-reliance. It also allows teachers to introduce, develop, extend or reinforce theoretical concepts, to develop scientific method, a critical awareness of experimental design and the ability to interpret data. It enables them to continue to promote the higher intellectual skills involved in scientific thinking when solving problems.

Not all the aims attributed to practical work need to be assessed, and it may not be possible to assess some of them in any valid or reliable way. However, if practical work is considered to be an essential activity, it should be assessed and the student's achievement incorporated as part of his or her final grade. A teacher's list of *teaching* aims for practical work may well be longer than a list of those which are to be *assessed,* but the assessor's judgement should be concerned with skills which are generally recognised to be important and which are also assessable. A consideration of the aims given above leads to the conclusion that the teacher, during the length of a course, *should* be more effective than a single terminal practical examination in judging the achievement of the aims.

Advantages of internal assessment

A higher reliability should be achieved by using internal assessment as it occurs on more occasions than a one-off practical examination. Ongoing internal assessment also means that more types of practical activity can be tested, giving a better sampling of biological content and practical skills. In particular, the practical *procedures* can be assessed, a difficult thing to do when dealing with large numbers of candidates at the end of a course. In addition, students can be assessed on practical work which forms part of a teaching sequence, and the stresses associated with a single examination are reduced or absent, as also are the failures due to accident or chance on a one-off occasion. Further, the teacher, being faced with the role of examiner, may well come to consider more carefully his or her teaching technique and any remedial action for individual students.

In spite of the difficulties and problems which it was thought that internal assessment would pose, there is a growing opinion among teachers and those concerned with public examinations that the teacher is in the best position to assess his or her own students and that the science schemes in action are working well. Achieving comparibility between the assessment of different teachers can be coped with by moderation. Student development during the course can also be accommodated in a flexible scheme and the teacher-student relationship seems to be as robust as ever.

As far as teaching techniques are concerned, internal assessment should be seen as a liberating influence. This may seem impossible to the new teacher, but teacher assessment does provide an opportunity to be released from the tedious repetition of certain exercises which prepared pupils for the traditional practical examination. Laboratory and field sessions can become more varied, meaningful and relevant to the course, and emphasis shifts from training to perform, to the development of basic scientific skills.

It must be admitted, however, that not all exercises may be used reliably for assessment, and thinking ahead is needed in order to satisfy both teaching and examination requirements. It is common at the start to place students in situations which are too sophisticated or open-ended. The average student cannot cope, which means that assessment cannot even begin, and laboratory space, staff, time and finance probably prohibit such assessments too. Whilst more demanding extension work may be used later in a course to discriminate between the more able candidates, simplicity is the overriding feature of all the guidance given here.

Aims

This book, then, aims to be a practical guide for teachers involved in the internal assessment of advanced biology courses. It gives ideas on the type of practical work which can be undertaken, and advice on laboratory techniques and methods of grading students. It also aims to reassure staff that there are many routes to assessment, all equally valid. Teachers should not be afraid of continuing to use well tried and successful procedures as well as introducing work devised by themselves. It has been the authors' experience in running in-service courses that teachers have much to gain from each other in designing practical work of worth, and that local consortia of teachers are a sound idea. This book is about internal assessment, but it is important to bear in mind that teacher-student contact is primarily about teaching and learning and not about assessment. Assessment should be secondary to the teaching aim. With planning and forethought this can certainly be so.

Chapter 2

ANALYSIS OF PRACTICAL SKILLS

It is important to define distinct practical abilities in order to facilitate the teacher's assessment. A clear list helps towards objectivity throughout each practical exercise as well as providing a means of quantifying the work of each student. In addition, and ultimately of greater importance, is the emphasis that this identification of skills places on the fundamental aspects of laboratory and field work. Hopefully a better balance of exercises is performed and consequently better practical biologists are trained.

Practical skills

Below, and in the following chapters, six separate practical skills or abilities are identified and the assessment of each commented on. Each is incorporated, one way or another, in the lists of abilities to be assessed which are provided by those GCE examining boards requiring teacher assessment in Advanced level biology (see Appendix 1).

The skills identified are as follows:

1 Manipulative skills

Under the heading of manipulative skills the manual dexterity of the student is tested in terms of his or her ability to:

handle chemicals and assemble apparatus
cut and prepare sections
use a hand lens and microscope at low and high magnifications
handle dissection instruments satisfactorily.

2 Following instructions during practical work

Assessment here marks the student's ability to complete an investigation in accordance with a specified procedure. It includes more than just the blind following of steps given, as it also involves an understanding of the instructions, enabling the student to make adjustments to the method if necessary.

3 Observation, identification, recording and interpretation

The assessment of these skills is based on the student's ability to recognise, identify and interpret biological material both microscopically and macroscopically. It also includes the clear and accurate recording of findings so that results can be understood by someone who did not see the original observation or investigation.

4 Presentation of experimental results with calculations
The student is assessed on his or her ability to select and / or implement the most appropriate method of recording the data collected (or, in some cases, given). In addition, grading is based on the relevance and accuracy of well-explained calculations, including elementary statistical analysis.

5 Interpretation of data
Under this heading, students are asked to analyse experimental results of both a qualitative and quantitative nature and to draw significant conclusions. Assessment will discriminate between those who appreciate the limitations of the data and the weaker candidates who, regardless of errors inherent in the procedure, make bold statements of fact based on theory work.

6 Experimental design
Whether the experiment which is designed by the student is also to be carried out, or whether it is a theoretical exercise only, the stages involved in the assessment remain the same. The student must show an ability to recognise a problem, formulate a hypothesis, devise a logical and timed work-plan and, choosing appropriate equipment and techniques with suitable controls, test the hypothesis. Finally, he or she should be able to choose a suitable method of presenting the results obtained and, by reference to the initial hypothesis, draw meaningful conclusions from them.

Assessment of these skills

Chapters 4 to 9 are devoted to the assessment of each of these six skills. They include a list of suitable practicals, plus three or four exercises in detail showing how they can be used in the laboratory for assessment.

For each exercise a worksheet is given, including instructions for the practical work as well as a series of structured questions. After the practical session the student will hand in drawings, graphs, written responses, records and so on for marking. The non-assessed work such as the writing up of method can then be completed in the student's own time. When assessment is complete and the material returned to and discussed with the student, he or she will be able to put together a complete record of the investigation and so build up a practical file.

In chapters 4 to 8, each investigation is made up of a number of sections. A list of equipment, resources and chemicals required by each student, all relatively cheap and easy to find, is given. The worksheet outlining the main steps of the procedure is written out in full under the heading 'Instructions given to students.' The follow-up includes instructions on how the observation and/or data should be recorded and processed. It gives structured questions which stimulate the student to analyse the practical procedure in terms of its errors and limitations, and to interpret the findings. Sometimes extension work is suggested.

Notes on assessment follow. Firstly a checklist, which is a summary list for the assessor, is given. This highlights the important steps in the exercise and emphasises them as separate from the practical details. Not all the items in the checklist are for assessment each time. By reference to the checklists from a series of exercises, however, teachers will soon see if there has been unnecessary repetition of skills tested or if there are gaps in the students' experience of techniques and abilities.

The major ability to be assessed by the practical exercise is then briefly considered, followed by a sample mark scheme for the assessment of written work or a set of criteria for the assessment of laboratory procedures. The marks awarded, or the scale of 1 – 10 associated with the criteria, will give a rank order and can be translated into the scale of marks required for the purpose of completing record cards for the examining board. Finally, other abilities which could be assessed by each exercise are considered.

Chapter 3

ASSESSMENT OF LABORATORY PROCEDURES AND OUTCOMES

Types of laboratory experience

What kinds of practical laboratory experience are suitable for assessment over the length of a course? A variety of activities can be used. Normal laboratory practical work which is done as part of the teaching of a particular topic is suitable, and the practicals detailed here would fit into the relevant teaching sequence in whatever way a particular teacher usually incorporates his or her practical work. Very many other practical investigations are possible, and further ideas are listed at the start of each chapter. The practical exercises may be of the illustrative type where known facts are confirmed, or investigatory, enquiry-based experiments presenting a new and more exacting problem. The usual types of practical work and laboratory experience which form part of the teacher's battery of techniques can, then, be used to give the ongoing assessment. Instead of assessing completely new work, a teacher may at times choose to repeat a previous practical or, preferably, some adaptation of a previous exercise using the same skill.

Besides normal laboratory practical work, there are other types of activity which can be used in assessment. A teacher demonstration, particularly where the practical involves the use of large or expensive apparatus, can be structured so that students in the class make their own observations and interpretations. An extension of this technique is to use data which have not been gathered at first hand by the class, but which have come from some other source, perhaps a previous year's class, a text book, a past examination paper, or a book of problems. The latter two sources have the advantage that questions on the data are already prepared and there are available examiners' reports indicating performance on the examination questions or books of answers to go with the problems.

On occasion an assessment will take the form of a school examination, particularly at the end of term or year. Project work and field work are also used for assessment, but the fact that they are long pieces of work does not mean they should be accorded a greater weighting than other practicals. The end product itself may not be what is considered the major part of the assessment, but rather the practical skills and thought processes which have been used during its production. In particular, experimental design or the planning of investigations is a skill which is usefully assessed using these two types of activity (see Chapter 9). Projects and field exercises need even greater forethought than do normal practicals before being integrated into assessment schemes because of their length, the complexity of the activities which go into them, the fact that they may be a group activity and because the teacher is inevitably greatly involved. However, the reverse side of the coin means that both projects and field work give a 'whole' approach to a biological problem, and allow integration of methods and content as well as student co-operation.

At the other end of the scale from the whole approach given by projects and field work are small-scale exercises consisting of a simple task assessing just one skill briefly and concisely. In assessment there is obviously a need for a clear specification of objectives, and several attempts have been made to produce a hierarchical list of objectives which would give even more guidance than the usual breakdown of skills. These objectives could then be assessed by short practicals. For example, manipulative skills may be subdivided as in Chapter 2 and a specific exercise to test each subskill could be devised: given a microscope, lamp and prepared slide the student could be observed as he sets up and focuses the microscope, followed by the viewing of his result. Similarly the ability to take readings, to put together a piece of apparatus, to follow a brief set of instructions or to carry out any other specified operation can be tested effectively as discrete abilities. All students have the same task, the same amount of time and the same apparatus, and the exercise is short and delineated. A collection of such items could be used in a single testing situation with a station approach where students watched by an assessor with a grading sheet, proceed round a laboratory and spend a few minutes at each station.

Such a hierarchy of practical skills which would allow, to quite a considerable extent, independent assessment of the specified qualities has been put forward by Whittaker (1974) and is reproduced with permission in Appendix 2. Although a categorisation of this nature is quite different from the current lists of assessed practical skills used by examining boards, it could be a useful aid in developing practical learning strategies.

Assessment

What is one actually assessing? The mechanics of the various schemes of internal assessment are provided by individual examining boards and, obviously, techniques of procedure and grading will vary from one board to another. However, some general points can be made and are worth emphasising.

A glance at the list of practical abilities given for assessment by an examining board shows that some are **procedures** and some are **outcomes.** These may be called **process** and **product.** The process skills in laboratory activities usually have to be observed in order to be assessed, and are probably assessable only by the teacher. The products have in the past formed the basis of the practical assessment made as a result of an external practical examination where it was assumed that the process by which the product was reached was the skill ostensibly being tested. For example, in leaving a labelled diagram of a microscope slide the student was assumed to have observed and inferred as indicated in the drawing; or in describing the procedure and end results of a practical test, the assumption could be made that she had indeed performed the test and observed its progress in the way detailed in her account. Without himself observing the process, however, the examiner could not know if this were true or if the student had arrived at her product in some other way, perhaps by calling on her theoretical knowledge after the failure of her practical procedures.

The **procedures,** then, are the activities done by the student during the practical work. They include such skills as manual dexterity and manipulative skills, the actual handling of equipment, materials and organisms; the use of biological techniques such as dissection; the use of instruments; the ability to follow instructions, understanding and adapting as necessary; observation and indentification; the taking and clear ordering of results; the practical procedures which go with experimental design, and so on. Even oral contributions to group discussion of a practical problem can be included here.

It is these procedures which are so difficult to assess during the course of a traditional practical examination. Given school-based teacher assessment, however, the task becomes possible. The activities must be assessed during the practical as the teacher observes the students at work and notes their progress. To this end the teacher can utilise a number of techniques. A schedule or **checklist** of points may be used. Such a schedule can range from an open-ended reference like those given as general criteria in Chapters 4 and 5, through to a precise list of critical actions to be noted, giving specific performance criteria against which a task is judged. Examples are whether or not a student heats the material in a test tube correctly and safely, uses a thermometer to take water temperature correctly, or strips a blood vessel cleanly and with care. With regard to, say, orderliness and tidiness of working, a scheme which involves marking a number of items on a three-point (0, 1 or 2) scale (for example, general tidiness, best use of bench space, relative positions of equipment, safe placing of dangerous or fragile pieces) could easily be devised and implemented.

Checklists give systematic, directed observation which is a great deal more useful in assessment than random, undirected observation. Those used in the individual practical tests administered as part of the work of the Assessment of Performance Unit, reported in *Science in Schools*, (1981, 1982), show very clearly the precision and exactness possible in the careful construction of such lists when scientific behaviours are being observed. While time-consuming to prepare, checklists are effective, particularly with small classes and for specific, discrete and easily observable operations. They also enable a teacher to pinpoint more readily student weaknesses, ineffective instructions, or problems with the worksheet. Such a scheme also allows for questioning of, or discussion with, the student, probing his or her insight and understanding. A mark for practical procedures may thus be built up during the practical by observing a pre-determined number of precise actions or activities.

While the use of a structured schedule is particularly appropriate for some types of practical skill such as observation, accuracy and planning, use of instruments and so on, it is not useful for all. For practical skills which seem to have a broader basis, a closely-defined checklist is impossible or not much use. Manual dexterity, for example, involving the rapid and confident completion of tasks, or the ability to work in a methodical fashion involving, say, correct sequencing of tasks, effective use of materials, good use of time, and ability to adapt instructions to one's own way of working, do not readily yield themselves to a detailed point-by-point analysis. In such circumstances an **impression mark** on a five- or ten-point scale is usually given, after obtaining an overall view of a student's work by observation during the practical.Impression marking on a particular occasion, or even after a period of time, interferes less with the course and does not create a feeling of artificial or excessive assessment. It is therefore a useful technique on the right occasions.

It is interesting to note that when Kelly and Lister (1969) compared two methods of assessing practical work in the then new Nuffield Advanced Biology scheme, they found that an impression mark awarded at the end of a term was comparable in accuracy with an assessment based on three centrally set practical exercises conducted during the term. The two methods of assessment which were compared represented the extremes of guidance which could be given to teachers. The practical exercises assessed were presented as actual tests, which were highly structured both with regard to experimental procedure and assessment. The situation was therefore different from the individual teacher assessment of ongoing practical work which is considered here, where a structured approach devised by the teacher for those occasions and abilities which warrant it should give more reliable results than an overall teacher impression mark.

Practical procedures, then, may be assessed in a number of different ways to suit the skill being assessed and the circumstances of the teacher. It is not really possible to assess more than one practical procedure on one and the same occasion, and certainly well nigh impossible to try to award impression marks for two skills during the same time.

The **outcomes,** or **products,** the results of the practical exercise, may be presented in a number of different ways. These include diagrams, tables, graphs, calculations, statistical analyses, written accounts, answers to questions, as well as practical end points such as a finished dissection, prepared slide, chromatogram or microbiological plate. It is not always necessary to assess an outcome on written work. Oral questioning and discussion can be used instead. As it is the written outcomes and final practical preparations which have traditionally been assessed in practical situations, their assessment poses few problems. Mark schemes or lists of points of one sort or another can be prepared in advance and applied to the material which is being assessed. Several such mark schemes are given in later chapters.

One point to be watched is to ensure that it is the student's own unaided work which is assessed. The practical session itself is obviously actively controlled and supervised by the teacher in charge who is intimately aware of what is going on. This does *not* mean carrying out the practical under examination conditions. If, however, the written-up work is handed in later for assessment, then clearly the teacher, while probably being able to guess accurately, is not in such a good position to pinpoint irregularities. Material for assessment must be collected at the end of the session, or some other foolproof arrangements made to obviate copying or outside help.

Assessing both procedure and outcome means that more than one ability can be assessed per practical, one during the exercise and one after. Care should be taken to see that students who have performed badly on the practical itself and thus have very little in the way of actual results to work on when it comes to the second stage are not penalised twice. In such cases, the interpretation stage can be performed on materials or results which are given to the student, and examples are described later in the book.

The schemes of different examining boards vary with regard to the specified list of abilities to be tested, the method of assessment, the criteria for the award of marks and so on. However, any assessment made by a teacher, whether involving a detailed and lengthy mark scheme or an impression mark on a one-to-five scale can easily be translated into the scale of marks required by the examining board, as well as the school's own grading system for internal records and transmission to students. For the examining board the important thing is probably to put the students into the correct rank order. If the teacher's actual marks are spot-on, all well and good. If the teacher is too lenient or too strict in comparison with most other teachers, the moderation procedure will correct standards.

Moderation

A teacher may judge his or her own students sensibly and reliably. But she may not. Apart from local colleagues, a teacher has probably little idea of how her own standards compare with others. Is she lenient or strict? Does she discriminate as well as others by using a similar range of marks with similar candidates; is she giving marks which are too high and too low or, alternatively, bunching the candidates together? Does she understand the criteria being used in assessment and therefore use her judgement soundly, producing the same rank order others would produce? In Schools Council *Examinations Bulletin 5* (1965) these three possible causes of poor teacher assessment are referred to as Standard, Discrimination and Conformity.

As standards will inevitably vary among teachers, it is essential to relate the teacher-assessed mark to some common standard, and the success of a scheme involving internal assessment depends on the **moderation.** Moderation is the process by which comparisons can be made between different groups of students assessed by different teachers in different schools with varying facilities, each teacher using his or her own set of laboratory practicals and applying the assessment techniques in the way which best suits his or her own situation. It is a way of equating the standards of individual teachers and the marks given with those of the whole student population entered for the same examination.

There are a number of possible methods of moderation of practical work.

1 A series of standard mini-tests of prescribed exercises
These could be provided by the examining board together with detailed objective mark schemes. Teachers would be expected to implement these tests and follow the scheme of marking to give a set of marks which would hopefully be equivalent for all candidates. Immediately, however, one can see the freedom and flexibility of the teacher's internal assessment being eroded, and the teaching permeated by formal, externally set assessment. And, with possibly well over a thousand teachers assessing a large number of students, such a method is not really possible without considerable and ongoing training.

2 Moderation by inspection
This is where the assessed work of all or of a sample of the students is inspected in order to compare it with others. The work produced could be re-marked by another assessor or moderator. Alternatively, the moderator could visit the school. However the work is inspected, the assessors or moderators should be able to assess student performance in laboratory procedures as well as the completed practical books, and they would therefore need to visit the school at the correct time and on a number of occasions. While such a method of working towards a common standard is useful within a school, or even between neighbouring schools, it is clearly not practicable to set up and monitor on a large scale. Even if there were enough moderators available, the method poses phenomenal problems of organisation and cost and is therefore impractical for a large-entry science examination.

3 A statistical method of moderation
Where a large number of candidates is involved, a statistical method is the most suitable. There must be a **moderating instrument** which is an examination component taken by all candidates. It gives other marks for all students against which the school assessment can be compared, and thus allows all to be judged on a common standard. The most common statistical method in use is scaling, where the students' performance on the test or examination designated as the moderating instrument is used to scale the teacher's marks for the same students in terms of standard and range of marks. As the school's rank order is not changed, any alteration to the teacher's marks results in a moving up or down of every student's mark.

What could a moderating instrument be? It *could* be a short practical test set and administered by the examining board. To re-institute such a test, however, after replacing a practical examination by teacher assessment, would be a backward step. It would cast doubt on the principles underlining internal assessment, perpetuate the problems of a practical examination and again reduce the flexibility and freedom of the teacher.

It is more usual to use as the moderating instrument part (or the whole) of the written examination. Such a common reference test should be chosen with care for its suitability as a moderating instrument. There are three points to bear in mind. It should be seen to be suitable by the teachers, that is, the objectives tested on the paper should relate to those being tested in the internal assessment. For example, the written paper could test a sub-set of the skills assessed by the teacher. Obviously, actual laboratory precedures cannot be observed on such a written paper, but questions can be asked such that only students who have performed the procedures can answer adequately. Questions can also be asked which test other skills associated with laboratory work and which form part of the teacher's assessment, such as data analysis and interpretation, hypothesis formation and the planning of investigations. The second point is that the moderating instrument should be made up of compulsory questions, set on a section of the syllabus which all candidates have studied so that all are in the same position. Thirdly, it should be capable of being marked objectively and reliably.

Other points

There are a number of points about continuous assessment for teachers to resolve for themselves, with whatever guidance is given by the examining board, and to keep consistently as policy during the course. Whether or not to tell the student in advance is one such point. Others include: how to make allowances in the assessment for any teaching which goes on during the practical, especially of help to individuals over and above that given to the class as a whole; if students are working in pairs or small groups how to ensure that each individual is assessed correctly; how and when to indicate their achievement on an assessment practical to the students; how to cope with absenteeism; at what stage in the course to start and end the assessments, and how to allow for student development over the length of the course; whether or not to use or discard marks for an assessed practical where things have gone wrong and most have performed equally badly. For teachers new to a scheme of internal assessment discussions with experienced colleagues on such points will guide the way to a strategy which suits their own situation. Other problems which may arise, such as persistent and deliberate absenteeism, transfer and repeat students, physical handicap, cheating or excessive help between members of the class should be covered in the instructions sent out by the examining board.

Chapter 4

MANIPULATIVE SKILLS

Under the heading of manipulative skills the manual dexterity of the student is tested in terms of his or her ability to:
handle chemicals and assemble apparatus
cut and prepare sections
use a hand lens and microscope at low and high magnifications
handle dissection instruments satisfactorily

Manipulative skills

The ability to handle and manipulate apparatus, organisms, chemicals and other materials is a basic skill in biology. Real success as a practical biologist can never be achieved if manual dexterity is not developed. Training and practice play a significant part in acquiring competence, and students will probably still be learning as they are assessed.

For assessment puposes there is no shortage of suitable practical exercises. Indeed, the problem may be in the selection of the most suitable exercises during the biology course. The final mark should cover the various stated components included under the heading of manipulative skills (such as dissection, section cutting, handling chemicals, assembling apparatus, and microscope work). The assessed practicals should therefore be chosen to give a representative sample of each skill, as well as being spread over the course to allow for individual development.

As manipulative skill is something which is demonstrated by a student in action, it must be watched and assessed during the practical session. The assessment techniques used by the teacher are discussed in Chapter 3. They may include observation of particular operations during the course of the practical (with or without a specific checklist of points to be watched), oral questioning of individual students, discussion with one or more students, and making notes on progress and achievements. The assessment may be continued after the end of the practical work to include a review of the completed task and the marking of the records made during the practical and the conclusions drawn from the results. If the particular skill being assessed is still fairly new to the students, some allowance will have to be made in the assessment of the direction and assistance given to individuals over and above that considered necessary for the class as a whole.

Practical exercises suitable for the assessment of manipulative skills

Dissection

A wide range of material is familiar to most teachers. Popular animals for dissection are mammal (rat, mouse), frog, fish, insects (locust, cockroach), earthworm, bird, mussel. Organs such as the heart or eye are also suitable.

Preparing material for examination

Sectioning, staining and mounting of plant parts	Maceration of plant material
Pollen grain squashes for pollen germination	Leaf epidermal strips for stomata
Chromosome preparations for mitosis and meiosis	
Flower dissection	Blood smears

Use of microscope

Examination of students' own preparations	Use of stage micrometer
Examination of prepared slides	Use of haemacytometer

Handling apparatus and chemicals

Very many investigations are possible, including:
Enzyme investigations

Water potential investigations

Physiology investigations involving the use of large pieces of apparatus such as spirometer, colorimeter, potometer, respirometer, gas burette

Microbiological techniques

Handling and counting *Drosophila*

Ongoing assessment of this nature during a practical necessitates giving careful thought in advance to produce a precise plan of assessment. The most suitable techniques for a particular practical should be decided on. Observation checklists, mark schemes, lists of questions and discussion points should be drawn up and sufficient copies prepared to free the teacher to give all his attention to gaining the assessment information he has decided on. Such assessment is a demanding and time-consuming job, especially with a large class. Ten students is probably the number above which one should think of bringing in a second assessor, splitting the class or cutting down on the ideal assessment to make it more manageable. Staggering the start of the practical enables the teacher to observe more easily critical points in the procedure for each student. If numbers, time or the nature of the practical are such that a detailed list of teacher activities is not possible or desirable, an overall impression mark can be given to each student at the end of the practical.

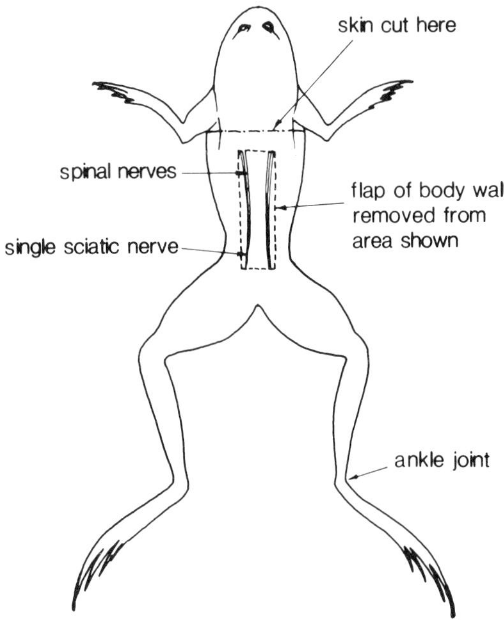

Fig 4.1 Dorsal view of the frog

INVESTIGATION 4.1
A frog dissection to display the muscles and nerves responsible for moving the foot

Equipment per student

1 fresh or preserved frog
Cork mat
Dissection dish and water
Pins
Dissection instruments

Instructions given to students

Dissect the hind limb of the frog so that you display the nerves and muscles which are responsible for moving the foot at the ankle joint. The procedure you should follow is given below.

1 Cut through the skin right around the body just behind the fore-limbs. Hold the skin with forceps and pull it back off the body and legs. You can now see the muscles.
2 Pin the frog onto the mat, *dorsal* side uppermost. Take hold, with forceps, of the very end of the vertebral column (urostyle) and lever it slightly upwards.
3 At the same time, snip through the body wall just below the forceps and make a cut forward on each side, leaving the nerves behind in the body.
4 Now cut across the forward end of the urostyle to remove a rectangular flap of body wall completely (see Fig 4.1).
5 You should be able to see, through the opening, several spinal nerves. Posteriorly these converge to form, on each side, the single sciatic nerve.
6 Trace this sciatic nerve through the thigh region by separating the muscles, firstly pulling them apart and then pinning them out. It is not advisable to cut through the muscles.
7 Finally, trace the sciatic nerve and its branches to the muscles which move the foot at the ankle joint. Dissect the muscles to make their outlines and attachments clear. Dissect one side only, but you could use the other side for practice.

Follow-up

Make a large, labelled drawing of your dissection of one side only. You are not expected to know the names of muscles or nerves, but label your drawing in terms of the position or function of the relevant parts.

Assessment

Checklist

1 Frog pinned out neatly and correctly.

2 Flap of body wall removed.

3 Sciatic nerve traced through the hind limb (see Fig. 4.2).

4 Muscles separated and pinned.

5 Dissection left tidily.

6 Correlation between dissection and labelled drawing.

It should be quite possible during the dissection to observe the students at work, question them individually, make notes on particular points and then, having inspected more closely the finished work, give a mark for the practical procedures involved in the exercise. A mark scheme will provide actual marks or a rank order.

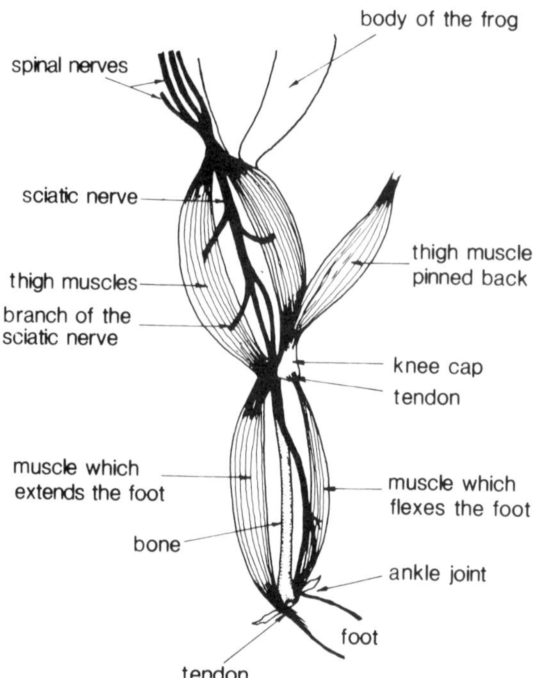

Fig 4.2 Dissection of the right hind limb of the frog

Mark scheme	
Works with confidence and care. The dissection is complete and accurate, the relevant bones, muscles and nerves are displayed fully	. . . 9–10
Able to apply the instructions adequately to the practical procedure but possibly works more slowly or less accurately and more errors made.	. . . 7–8
Shows basic understanding of techniques involved and the display is just adequate. Possibly untidy work and lacking in finish.	. . . 5–6
Limited in terms of technique and facility. Many careless mistakes and dissection incomplete.	. . . 3–4
Very poor ability in handling instruments and material. Very little attempt to complete the exercise and many parts damaged or destroyed.	. . . 1–2

Other abilities which could be assessed

This provides a good exercise if frogs are available from having been the subject of other dissections or practical work, as the structures investigated in 4.1 will probably still be intact and maximum use can be made of the animal. Frogs should not be killed solely for the assessment: to do so would be a needless waste of life.

Observation, recording and interpretation

The exercise gives a good assessment in that students have to follow instructions carefully in an unfamiliar dissection. They then have to observe accurately and apply general biological principles to interpret the features displayed.

By using a detailed mark scheme with the drawing and its labels, an assessment may be made of the student's ability to observe, recognise and record the salient points. In addition to allocating marks for presentation, scale, drawing and identification of the relevant bones, nerves and muscles, credit should also be given for the relationship between the drawing as a whole and the dissection as displayed by the student at the end of the practical.

INVESTIGATION 4.2
The sensitivity of the Benedict's test

Equipment per student

20 cm^3 10% glucose solution
25 cm^3 Benedict's solution
Distilled water
10 cm^3 graduated pipette or syringe
5 test tubes and rack
2 x 250 cm^3 beakers
Tripod and gauze
Bunsen burner
Wax pencil
Coloured pencils

Instructions given to students

1 Set up a water bath containing boiling water.
2 Label five test tubes 1 to 5 with a wax pencil near the top of each tube.
3 Pipette 10 cm^3 10% glucose solution into tube 1.
4 Transfer 1 cm^3 of this solution from tube 1 to tube 2. Using the pipette, add distilled water so as to prepare, in tube 2, a 1% glucose solution. Mix well.
5 Using firstly 1 cm^3 of the solution from tube 2, go on to prepare three more glucose solutions systematically:
tube 3: 0.1%
tube 4: 0.01%
tube 5: 0.001%
6 Finally, adjust so that all the tubes contain an equal volume of liquid.
7 Pipette 5 cm^3 Benedict's solution into each test tube.
8 Transfer all five tubes into the boiling water bath and leave for three minutes.
9 Remove and return them to the rack.
10 Compare the colours in each tube and tabulate the results using coloured pencils.

Follow-up

1 Give full details of how you prepared the serial dilution of glucose from 10% to 0.001%.
2 Why did you use a water bath rather than heat the tubes directly over a Bunsen burner? Give three possible reasons.
3 From your results, what would appear to be the least concentration of glucose that Benedict's solution will detect?
4 How could you extend the investigation and make it more accurate?

Assessment

Checklist

1 Apparatus handled skilfully and organised methodically.

2 Instructions followed and procedure carried out with care.

3 Serial dilution prepared accurately and with understanding of method.

4 Results logically and neatly tabulated.

5 Results interpreted and conclusion drawn with a clear understanding of the limitations of the method used.

The handling of the apparatus and the precision employed in the preparation of the serial dilution can be judged both during the practical and from the written account of the method used. A mark scheme will provide actual marks or a rank order.

Mark scheme

A meticulous and accurate worker who fully comprehends the methods employed.	. . . 9–10
A less confident and perhaps slower approach, but competent in completing the procedure.	. . . 7–8
A clumsier and less thoughtful approach, such that accuracy of technique is threatened.	. . . 5–6
The procedure is full of errors, both in method employed and basic practical techniques.	. . . 3–4
Little understanding of the experimental method shown and even basic instructions not executed: totally disorganised and very careless.	. . . 1–2

Other abilities which could be assessed

Observation, recording and interpretation

The best students will construct a logical table of results. They will record the colour changes accurately and interpret the visual data meaningfully whilst appreciating the limitations of the rather crude technique.

A mark scheme can be applied to the recording of results and the conclusions drawn.

Experimental design

The suggestions for refinements, modifications and extensions to the initial investigation may be done as a written account or may form the basis of an additional practical exercise for students. In either case the assessment of planning ability is possible.

INVESTIGATION 4.3
The distribution and nature of the storage products present in a lemon fruit

Equipment per student

Half lemon
Iodine in potassium iodide solution
Sudan III
Benedict's solution
Glycerine
Distilled water
Filter paper
White tile
Teat pipette
Watch glass
Bunsen burner
Microscope with light source
Microscope slides
Cover slips
Dissection instruments

Instructions given to students

1 Cut a thin section through the outer **peel** of the fruit at right angles to the surface of the fruit. It may be easier to remove a narrow strip of peel first. Mount the section on a slide and examine under the microscope. Make an annotated diagram to show structure, but do not draw cells.

2 Now test the peel for storage materials and cell inclusions as follows.

   a With iodine: place a new dry section of peel on a microscope slide and add several drops of iodine solution. Leave for 5 minutes. Remove the stain, add a drop of glycerine and mount under a cover slip. Record the result of the test and note the distribution of material present.

   b With Sudan III: repeat the test above, but using Sudan III reagent instead of iodine. Record the result and distribution of material.

   c With Benedict's solution: take a small piece of peel and place it at one end of a microscope slide. Chop it finely with a scalpel, add a drop of water and warm gently over the Bunsen burner. Add more water if necessary. Remove a drop of the solution with a pipette and transfer it to one end of a clean microscope slide. Add one drop of Benedict's solution, stir with a needle and then heat the slide until the solution boils. Record the result of the test.

3 Take a **seed** and with a scalpel and needle remove the outer seed coat. Now cut a thin transverse section through the fleshy tissue of the seed. Examine it under the microscope and describe briefly what you can see of its structure. Do not make a drawing.

4 Test the fleshy tissue of the seed for storage materials in exactly the wame way as for the peel, using iodine, Sudan III and Benedict's solution. Record all results.
5 Tease out a small portion of the JUICE-CONTAINING TISSUE. Examine it under the microscope and describe briefly what you can see of its structure. Do not make a drawing.
6 Now test the juice-containing tissue for storage materials, using iodine and Sudan III, in exactly the same way as for the peel. With Benedict's solution, however, place a drop of juice on one end of a microscope slide and add one drop of Benedict's solution. Gently heat the slide until the solution boils. Record all results.

Follow-up

What conclusions can you make about the nature and distribution of cell inclusions and storage materials present in the peel, the seed and the juice-containing tissue? Comment on the probable biological significance of these materials.

Assessment

Checklist

1 Sections cut and other materials removed and mounted satisfactorily.
2 Mounting, staining and microscope examination completed skilfully.
3 All materials, chemicals and equipment handled with care and precision.
4 Drawings, descriptions and tabulation of food test results presented clearly.
5 Conclusions drawn in relation to the food tests made.
6 Nature and distribution of materials present interpreted in terms of the function of a fruit and seeds, and with reference to method of dispersal and process of germination.

This particular exercise provides a very thorough and exacting test of manual dexterity. It involves most of the specified aspects of the skill: section cutting, handling equipment and chemicals, and the use of the microscope. It is a lengthy task and there is ample opportunity for the teacher to watch students at work and to note their strengths and weaknesses. By reference to a mark scheme, the students can be ranked or marks awarded.

Figs 4.3 and 4.4 overleaf show what will be observed by the student.

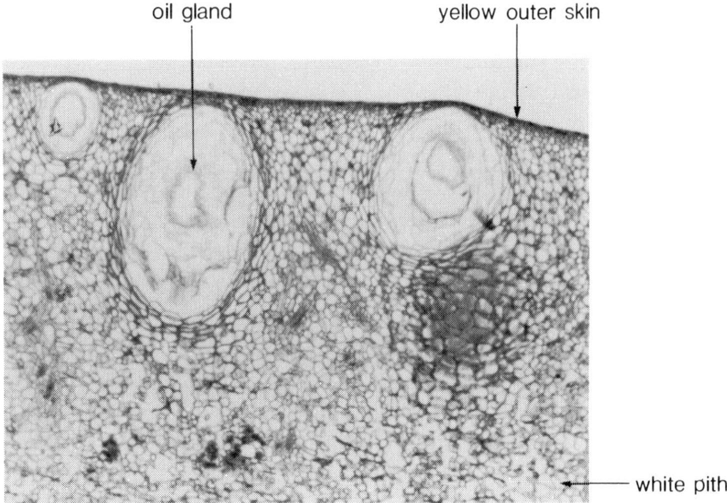

Fig 4.3 Citrus fruit: VS skin (×25)

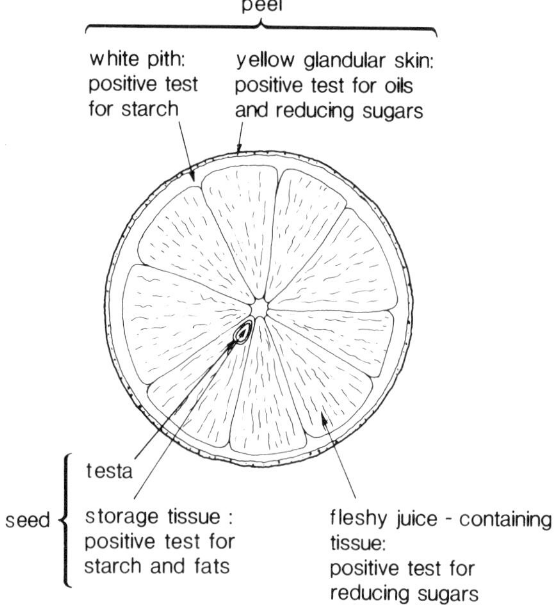

Fig 4.4 TS lemon fruit to show distribution of food materials

Mark scheme

Sections and materials prepared for microscope examination with skill and clarity. Works through the instructions confidently and smoothly, and handles all equipment with precision and thoughtful organisation. . . . 9–10

Some competence in cutting sections and carrying out the basic techniques displayed. However, possibly works more slowly or with less satisfactory results. . . . 7–8

Finds more difficulty in executing some aspects of the skill, eg mounting and staining. Nevertheless completes the task with some success and shows concern for bench organisation and safety. . . . 5–6

Experiences real problems in carrying out basic procedures. Has only limited ability in preparing sections and mounting material. Careless attitude to the use of the Bunsen burner and/or microscope, resulting in many errors. . . . 3–4

A clumsy and untidy approach. Sections are not produced at all or are very poor in quality. Chemicals and apparatus handled in a careless and thoughtless manner. Not one aspect of the exercise satisfactorily completed. . . . 1–2

Other abilities which could be assessed

Observation and recording

For each part of the exercise the student is asked either to draw or describe the material prepared for examination under the microscope. Assessment of his or her powers of observation may be made by using a mark scheme with the drawings or descriptions and with the records of the results of the food tests.

Interpretation of data

The laboratory part of this investigation is time-consuming, and individual results will be variable depending on the success of techniques employed in the early stages. It may be more realistic to assess the interpretation of data at a subsequent session when the findings from the class as a whole can be pooled and discussed. Individual students can then spend time drawing their own conclusions. In this way everyone in the class, however poor their practical skills, starts being assessed on the interpretation of data on an equal footing. Furthermore, the exercise becomes more productive as more data are available.

Chapter 5

FOLLOWING INSTRUCTIONS

Assessment here marks the student's ability to complete an investigation in accordance with a specified procedure. It includes more than just the blind following of steps given, as it also involves an understanding of the instructions, enabling the student to make adjustments to the method if necessary.

Following instructions

The ability to follow instructions for experiments and investigations and to perform specified procedures during practical work is a fundamental skill. It is one which is probably included as part of an overall umbrella title such as the carrying out of experimental investigations and applying experimental techniques, but which can be tested separately.

Almost all practical exercises can be used for this assessment and, like the assessment of manipulative skills, it will normally be done whilst the class is observed at work. In the same way as for manipulative skills, prior thought must be given to the assessment technique to be used and checklists or mark schedules prepared in advance. It is very difficult to assess at one and the same time two skills (such as those given in this and the previous chapter) which require surveillance with swift and efficient movement of the assessor round the class, and teachers are advised against it.

As students are being assessed on their ability to deal with instructions, it follows that the practical work being tackled is probably something new or possibly a different adaptation of a previous approach. Individual students may require help and guidance at different stages and to differing degrees. Help should be given where needed so that laboratory sessions are as instructive and productive as possible. Allowance must be made for extra help given to an individual, and a good mark scheme will cover this.

The unpredictability of biological material will mean that sometimes the exact and rigid following of instructions is not possible or indeed desirable. Mark schemes should discriminate between the better student who will show intelligent comprehension of the schedule and so be able to adapt and adjust procedures if necessary, and the less able candidate who will be lost if all does not go according to plan. The weakest students of all will, of course, have problems even with the basic routine.

Practical exercises suitable for the assessment of following instructions

Any procedure involving a sequence of steps would be ideal for assessment purposes, and examples appear in other chapters as well as this one. Additional practical areas include the following:

Microbiological investigations
Basic techniques of media preparation and plating

Culturing of bacteria from air and soil samples, milk

Effect of antibiotics, temperature on bacterial growth

Effect of pH, temperature, oxygen, sugar concentration on growth of yeast

Enzyme investigations
Less common examples include phosphatase from mung beans, phosphorylase from potatoes, dehydrogenase in yeast, urease.

Enzyme properties studied can include substrate specifity, effect of enzyme or substrate concentration, action of inhibitors

Locust, earthworm and small mammal guts, as well as germinating seeds, may be used to investigate location and activity of digestive enzymes.

Plant hormone work
Effect of light on coleoptiles

Effect of gravity on root and shoot growth

Use of the clinostat

Application of auxin to coleoptiles

Food tests
Identification of unknown solutions

Food reserves in plant organs

Water relations
Determination of water potential using change of mass, length

Water uptake and loss, use of potometer

Water culture experiments

Photosynthesis and respiration experiments

INVESTIGATION 5.1
The preparation of a root tip squash to demonstrate stages of mitosis

Equipment per student

Broad bean seeds: soak and allow to germinate for 7 days between damp cotton wool at room temperature
10 cm^3 50 : 50 concentrated hydrochloric acid and absolute alcohol
10 cm^3 45% glacial acetic acid
100 cm^3 70% alcohol
Acetic orcein in a dropping bottle
Microscope with light source
Stop clock
Scissors
Watch glasses with glass covers

Forceps
Mounted needle
Teat pipette
Scalpel
Beaker
Glass rod
Microscope slides
Cover slips
Labels
Lens tissue
Filter paper

Instructions given to students

1 Cut off the root tip, about 2 mm in length, from each seedling. Using a mounted needle, place the root tips in a covered watch glass containing a mixture of concentrated hydrochloric acid and absolute alcohol in equal volumes. Leave for 5 to 10 minutes.
2 Transfer the root tips to another covered watch glass containing 45% glacial acetic acid. Leave for 5 minutes.
3 Place a root tip on a microscope slide which has been kept in 70% alcohol and cleaned with lens tissue. Cover the tip with a drop of acetic orcein and crush thoroughly for 2 to 3 minutes with a glass rod. Do not allow the material to dry up.
4 Place a cover slip on the slide, tap all over the cover slip, and gently squash the preparation by pressing down on the cover slip through layers of filter paper. Label the slide.
5 Examine under the microscope and identify the main stages of mitosis.

Follow-up

Find one clear stage of mitosis from your preparation. Identify it and make a large, well-labelled diagram to show the detail.

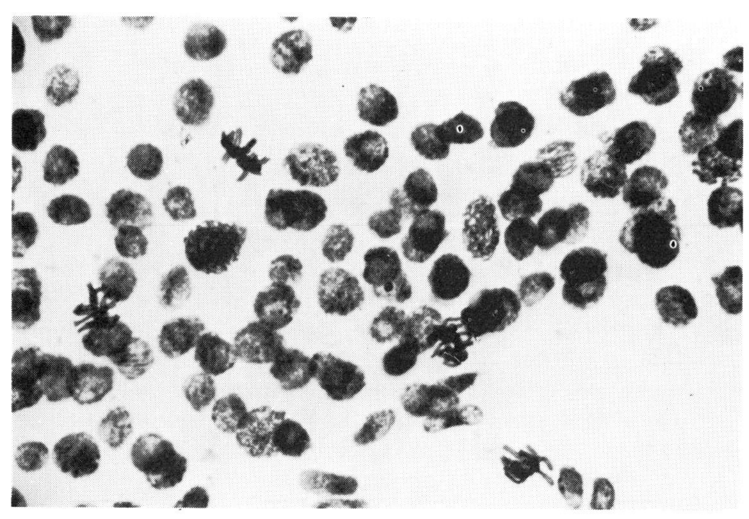

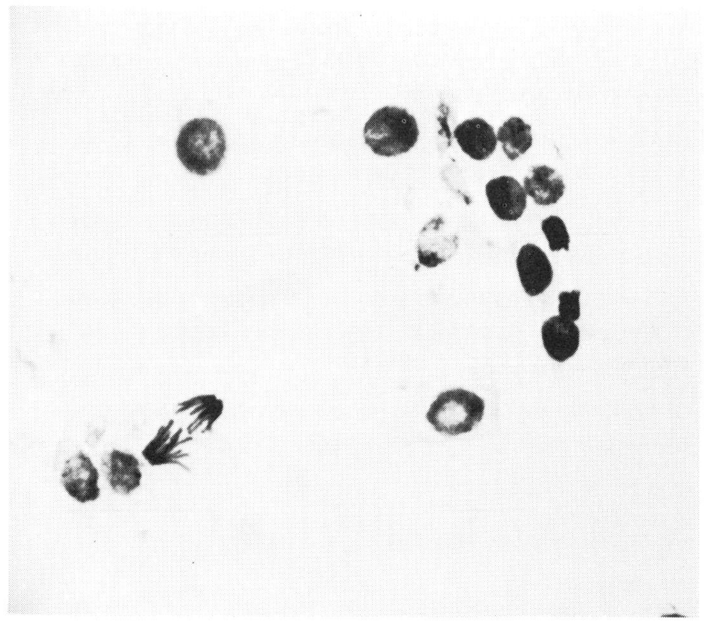

Fig 5.1(a), Fig 5.1 (b) *Vicia* root tip squash to show stages in mitosis (×250)

Assessment

Checklist

1 Follows instructions accurately and in sequence.

2 Handles chemicals, materials and apparatus with care.

3 Handles microscope correctly, using correct level of magnification.

4 Identifies, draws and labels one stage of mitosis. Fig 5.1 shows *Vicia* root tip squashes.

This ability to follow instructions will be assessed by a mark given from the close observation of students during the investigation. The practical procedure is quick and effective, and several preparations may be made during the course of a lesson. The more able students will adapt the technique and/or repeat to obtain better results so that a rank order should not be difficult to obtain.

Mark scheme

Follows instructions very carefully. Works methodically and thoughtfully at all times. Able to see errors made and immediately adapts or repeats procedure to correct them.	. . . 9–10
Follows instructions with care but shows less understanding of the techniques. Nevertheless, competent and thorough.	. . . 7–8
Able to follow the instructions but shows little discernment. Some help probably required if results are not initially satisfactory.	. . . 5–6
Careless in reading and following instructions so that errors arise which may not even be appreciated. Help certainly needed to complete the exercise.	. . . 3–4
Unable to complete the procedure at all and much assistance needed at all stages. Fails to respond successfully to advice given.	. . . 1–2

Other abilities which could be assessed

Manipulative skills

It would be inadvisable to assess manipulative ability at the same time as the skill to follow a practical procedure, since both necessitate close observation of procedures. However, the handling of the material, chemicals and microscope requires considerable manual dexterity, and the exercise could easily be used to assess these skills.

Observation, recording and interpretation

The students' diagrams are assessed by comparing the record made on paper with the preparation under the microscope. Marks will be given for the accuracy of the relationship between the two, and a separate mark scheme is applied to the drawing with title and labels. It is essential at the end of the practical to provide suitable material, probably prepared slides, for unsuccessful students. Fig. 5.2 shows stages in mitosis.

Since laboratory preparation of root tip squashes with identification and diagrams of all stages cannot normally be completed in the usual time available, a subsequent practical session could be devoted to the study of prepared slides of mitosis stages. This again provides an excellent opportunity to assess powers of observation and recording ability.

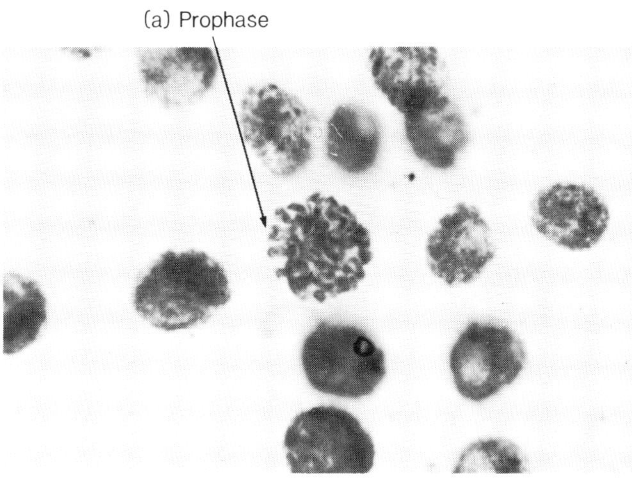

Fig 5.2(a) Prophase (×400)

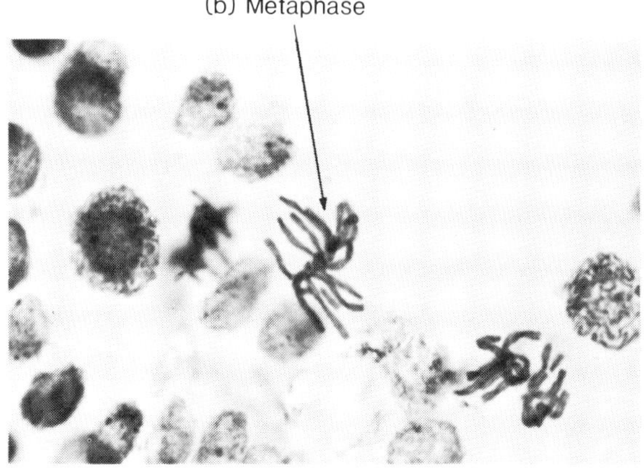

Fig 5.2(b) Metaphase (×400)

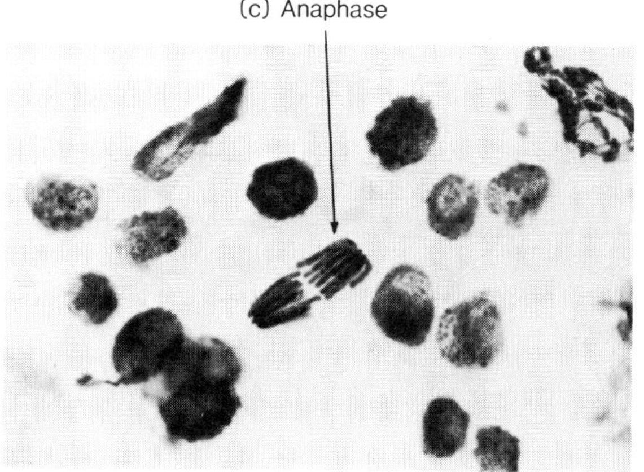

Fig 5.2(c) Anaphase (×400)

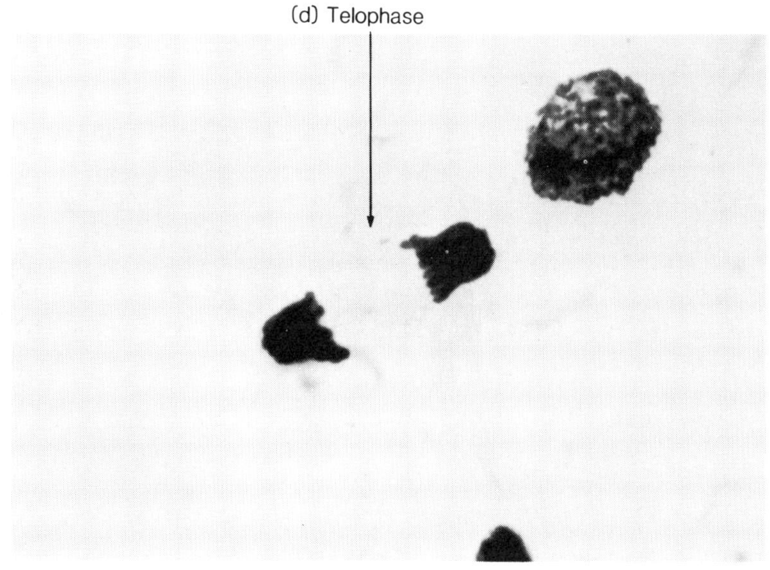

Fig 5.2(d) Telophase (×400)

INVESTIGATION 5.2
The effect of pH on invertase activity

Equipment per student

1.5 cm^3 invertase concentrate (kept in refrigerator)
7 Clinistix reagent strips
50 cm^3 4% sucrose solution
M/10 hydrochloric acid
M/10 sodium carbonate
Universal indicator solution
Distilled water
Universal indicator paper and chart
Stop clock
Wax pencil
7 test tubes and rack
2 cm^3 syringe
5 cm^3 syringe
Watch glass
Teat pipette

Instructions given to students

This experiment uses invertase, an enzyme obtained from yeast which catalyses the hydrolysis of sucrose to glucose. The presence of glucose is detected using a Clinistix strip, the colour of which changes from pink to dark blue when the solution contains 0.25% or more glucose.

1 Dilute the invertase concentrate in a watch glass by adding 6.0 cm^3 distilled water to 1.5 cm^3 invertase.
2 Place 1 cm^3 diluted invertase into each of six test tubes labelled 1 – 6.
3 Put 1 cm^3 of distilled water in a separate test tube labelled 7.
4 To all seven tubes add 3 drops of universal indicator solution.
5 To each add a few drops of M/10 hydrochloric acid and/or M/10 sodium carbonate solution in order to produce the following pH values indicated by specific colours.

tube 1	bright red	pH 3.0
tube 2	red	pH 4.0
tube 3	orange	pH 5.0
tube 4	yellow	pH 6.0
tube 5	green	pH 7.0
tube 6	blue	pH 8.0
tube 7	blue	pH 8.0

Work as closely as you can to this table but, in any case, record colours and pH values in each tube carefully.

6 Add to each tube 5 cm^3 of the 4% sucrose solution. *Immediately* put a Clinistix strip into each tube and note the time.
7 For each tube record the time for the reagent to turn from pink to blue. It is most important that you decide on a standard end point.
8 Tabulate results and record all observations.

Follow-up

1 Since the colour change occurs when the solution contains at least 0.25% glucose, the rate of reaction may be calculated as:

$$\text{Rate of glucose formation} = \frac{0.25 \times 60}{\text{Time(min)}}\ \text{g h}^{-1}$$

For each tube calculate the rate of glucose formation.

2 Tabulate these results with the appropriate pH values.
3 Plot a graph of rate of reaction against pH for tubes 1–6.
4 Estimate the pH optimum for invertase activity.
5 *a* Account for the result obtained in tube 7.
b By taking into consideration the result from tube 7, draw a modified sketch graph of the relationship between pH and invertase activity.
c What was the purpose of tube 7?
d In view of the findings from tube 7, suggest extension work to the practical procedure.

Assessment

Checklist

1 Apparatus and chemicals handled carefully and tubes set up precisely according to instructions.
2 Standard end point made clear and consistency observed.
3 Results tabulated logically.
4 Calculations of rate made and results tabulated.
5 Graph plotted and optimum pH information read from it. (This will be in the range 4.0 – 5.0.)
6 Appreciation of use of tube 7 as a control in view of instability of sucrose in the alkaline range and thus the formation of glucose in the absence of enzyme catalysis at pH 8.
7 Experimental design or modification of technique to include sucrose only controls at each pH value.

This particular investigation has advantages in that results are obtained at room temperature within 10 to 20 minutes. In addition, the enzyme and techniques employed are less familiar and so provide something of a challenge to the student.

The preparation of the tubes requires very careful consideration and organisation. Reasonable results will only be achieved if the procedure is followed through methodically and thoughtfully. The students can be graded on a 1–10 scale by teacher observation during the experiment.

Mark scheme

Efficient and knowledgeable approach to task. Logical and well organised preparation of tube contents. Confident about the end point. Bench space neat. . . . 9–10

Apparatus and chemicals handled correctly, although with a more tentative approach. Investigation completed, although with less skill. . . . 7–8

Generally able to follow instructions, but some assistance required in setting up the colour range and/or in determining end points. . . . 5–6

A more confused approach, with bench work careless and untidy. Inaccuracies in preparing the tubes and lack of comprehension in recording results. . . . 3–4

A failure to understand the instructions and follow them resulting in the tubes being unsatisfactorily prepared. Much assistance required; even then, procedure not completed. . . . 1–2

Other abilities which could be assessed

Presentation of results and calculations from data

Both during the practical procedure and when processing data, it is important that information and results are clearly and logically tabulated. The students' efforts to do this, along with accurate calculations of rate of reaction and good graph construction, will be graded from written work collected after the laboratory session. Students who have been unsuccessful in the experimental method and have no reasonable data to work on should use data from other sources so that they can be assessed in this skill.

Interpretation of data

The structured questions in the follow-up work lead the student to determine the pH optimum and to discuss the value of including a control tube. The application of a mark scheme to the written responses will enable data interpretation to be assessed.

Experimental design

The best students will appreciate the limitations of the procedure and techniques involved and will be able to suggest modifications and extensions. This part of the exercise may, however, be more productive for all if it is based on a later class discussion. Assessment can then be made either during the discussion by an impression mark or from written work afterwards.

INVESTIGATION 5.3
The estimation of metabolic rate of an organism using a simple respirometer

Equipment per student

Manometer fluid
10% potassium hydroxide solution
Components of the respirometer (see Fig 5.3)
Water bath with thermostat control
Thermometer
Living organisms such as soaked seeds, blowfly larvae
Rubber tubing
Watch glass
Stop clock

Instructions given to students

1 Set a water bath at 20°C.
2 Label two boiling tubes A and B (see Fig 5.3). Into A pour 5 cm³ 10% potassium hydroxide solution. Do not let it run down the side. Put in a small roll of filter paper to act as a wick.
3 Pour into B about 14 cm³ water. This reduces the gas volume in B to make it equivalent to that in A.

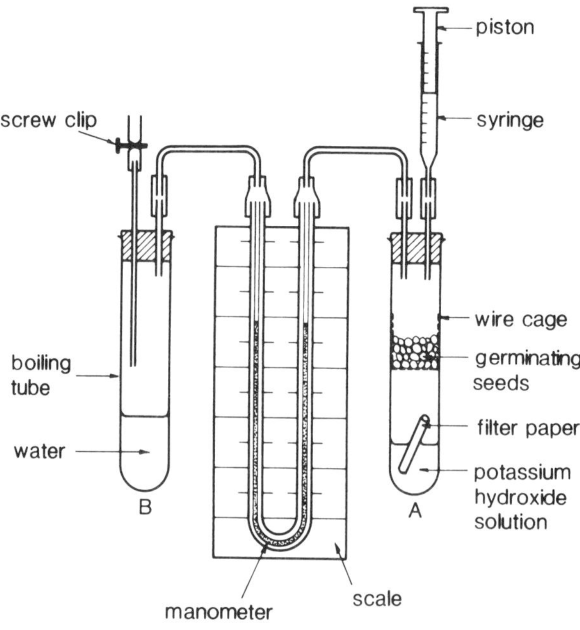

Fig 5.3 A simple respirometer

4 Fill the wire cage with a known weight of germinating seeds. Push the cage into A so that a bung may be inserted, but keep the cage well clear of the solution.
5 Insert a bung into A but do *not* attach the syringe. Insert a bung into B and keep the screw clip open.
6 Filling the manometer: Pour the coloured oil into a watch glass. Attach a piece of rubber tubing to one end of the U-tube and invert it. Dip the open end of the U-tube into the oil and suck gently. When a thread of oil has just reached the first bend of the tube, stop, lift the end out of the oil and turn the U-tube the right way up.
7 Connect the filled manometer tube to the boiling tubes, taking care not to force oil out of the U-tube.
8 Set up the apparatus so that the boiling tubes are in the water bath but the manometer is outside. Fix the manometer vertically to the side of the water bath with sticky tape.
9 Leave the apparatus for 5 minutes to equilibrate.
10 Pull the piston of the syringe halfway out and gently ease it into the rubber tubing of A. If the manometer fluid is displaced, bring it back to a central position by easing out the syringe piston.
11 Finally, close the screw clip on the rubber tubing on tube B. Again, adjust the manometer fluid level if necessary using the syringe.
12 Note and record the following *immediately*:
   a Exact position of meniscus in U-tube (both sides).
   b Exact position of the piston in the syringe. Take a reading.
   c The time.
   d The temperature of the water bath.
13 Allow the experiment to run for at least 30 minutes.
14 During that time, as the oil meniscus level reaches the bend in the U-tube, record the new levels. Then restore the oil to its original position using the piston of the syringe. Repeat as necessary, recording all results.
15 Finally restore the manometer fluid to its original level and take a final reading on the syringe. Note the time.

Follow-up

1 Record the following results.
   a Weight of the organism used.
   b Temperature of the water bath.
   c Time interval for experiment to run.
   d Initial reading on syringe.
   e Final reading on syringe.
2 Calculate the volume of oxygen absorbed during the experiment in cm^3.
3 Now determine the metabolic rate of the organism expressed as the volume of oxygen absorbed in $cm^3\ h^{-1}\ g^{-1}$ at the temperature used.

4 What control could you use for the two respirometers? Explain fully.

5 How could you use this apparatus and extend and modify the procedure to measure the effect of temperature on respiratory or metabolic rate? How could the temperature coefficient (Q_{10}) be calculated?

6 It is also possible to use this apparatus to estimate the RQ value of the respiring material.
   - *a* Repeat the experiment steps 1–15 as outlined previously, but place 5 cm^3 water in tube A instead of the potassium hydroxide.
   - *b* Again, record the time for the experiment, and the initial and final readings on the syringe.
   - *c* By noting the direction of movement of meniscus levels in the U-tube and change in position of piston in the syringe, calculate the new total volume of gas evolved or consumed.
   - *d* Knowing from the previous experiment the volume of oxygen absorbed in $cm^3\ h^{-1}\ g^{-1}$ of material used, calculate the volume of carbon dioxide evolved in $cm^3\ h^{-1}\ g^{-1}$.
   - *e* Finally calculate the RQ value for the respiring material.

7 These figures (converted to NTP) were obtained with respiring peas in the respirometer:

   Volume of oxygen consumed: 8.7 cm^3
   Volume of carbon dioxide produced extra to amount of oxygen absorbed: 0.8 cm^3

   - *a* Calculate the RQ value.
   - *b* What food was being used as the respiratory substrate?

Assessment

Checklist

1 Assembles apparatus with precision and skill.

2 Follows experimental instructions methodically and thoughtfully, and is able to adjust procedure if necessary.

3 Records results clearly and meaningfully.

4 Comprehends the theory behind the working of the apparatus and therefore the significance of the results obtained.

5 Completes calculations accurately and draws relevant conclusions.

6 Shows an understanding of the limitations of the apparatus, sources of error, need for controls and ways in which procedure can be extended for further investigations.

This type of apparatus and experimental procedure always seems to be problematical, and valid results may not be obtained. However, there is much to be gained by first-hand experience of the assembling of the apparatus and systematic completion of the procedure. For a start, theoretical data will be more readily understood and interpreted.

Assessment, then, of the students' ability to follow the experimental instructions regardless of actual results obtained can be done. Even though students will probably be working in pairs or small groups, which will make grading each individual's performance difficult, it should be possible to produce a rank order.

Mark scheme

Confident and business-like approach, well organised procedure throughout. Thoughtful at all stages and able to modify practical if necessary.	. . . 9–10
Careful and methodical execution of instructions, but shows less understanding of purpose of techniques. However, completes practical satisfactorily.	. . . 7–8
May need help in assembling apparatus. Works more slowly and carelessly, but with help can complete procedure.	. . . 5–6
Much assistance needed at all times. Fails to see the difficulties and problems. However, remains involved and determined to see the exercise through.	. . . 3–4
Very thoughtless and clumsy approach, with little or no dedication to the task.	. . . 1–2

Other abilities which could be assessed

Manipulative skills
Clearly the handling of chemicals and the components of the respirometer, together with the manipulation involved during the practical, test the manual dexterity of students. Marking of manipulative ability could take place during the session, but should not be confused with the ability to follow the experimental instructions.

Presentation of results and calculations
Experience has shown that results obtained are not always reliable, and of course if any comparisons are to be made, volumes of gases need to be converted to NTP. However, theoretical calculations of metabolic rates and RQ values will be more meaningful when pupils have collected and recorded data for themselves. If the data collected do prove to be unusable, second-hand data could be distributed and the calculations performed on reliable figures.

Experimental design
The student, having completed the basic routine, will be in a better position to consider other aspects of the procedure, such as sources of error and the need for controls. Extension work to involve estimates of temperature coefficient can also be considered, either theoretically or practically.

Chapter 6

OBSERVATION, RECORDING AND INTERPRETATION

The assessment of the skills of observation and interpretation is based on the student's ability to recognise, identify and interpret biological material both microscopically and macroscopically. It also includes the clear and accurate recording of findings so that results can be understood by someone who did not see the original observation or investigation.

Observation, identification, recording and interpretation

While observation, identification and recording are activities which take place in the laboratory, the assessment of abilities under this heading will probably involve the use of the actual records produced by the students. Such records may take the form of labelled and annotated diagrams, interpretations and explanations given as answers to questions, calculations, tables or graphs. The details of the written record itself can be marked after the laboratory session. During or at the end of the session, however, the assessor will probably need to check the relationship between the biological material observed or prepared by the student and the record left behind.

A well-constructed and detailed mark scheme should be used with the written record so that each student's work is assessed in the same way. The mark scheme can later be converted into the required grades or marks.

Practical exercises suitable for the assessment of observation, recording and interpretation

Plant morphology

Relationship between structure and functions of roots and shoots	Comparisons
Flower structure, comparative, pollination	Adaptations

Animal morphology

Relationship between dissection and structures identified	Skeleton work

Insect mouthparts and legs: comparative studies	Histological investigations
Ecology	
Distribution, succession	Use of flora and keys to identify and classify
Population studies	
Animal behaviour	
Comparison of related forms with reference to environment and mode of life	
Analysis of photographs, photomicrographs and electron micrographs	
Mitosis and meiosis	Cell structure
Plant and animal anatomy	Embryology and developmental stages
Biochemistry	
Food tests	Chromatography
Serial dilutions	Enzyme reactions

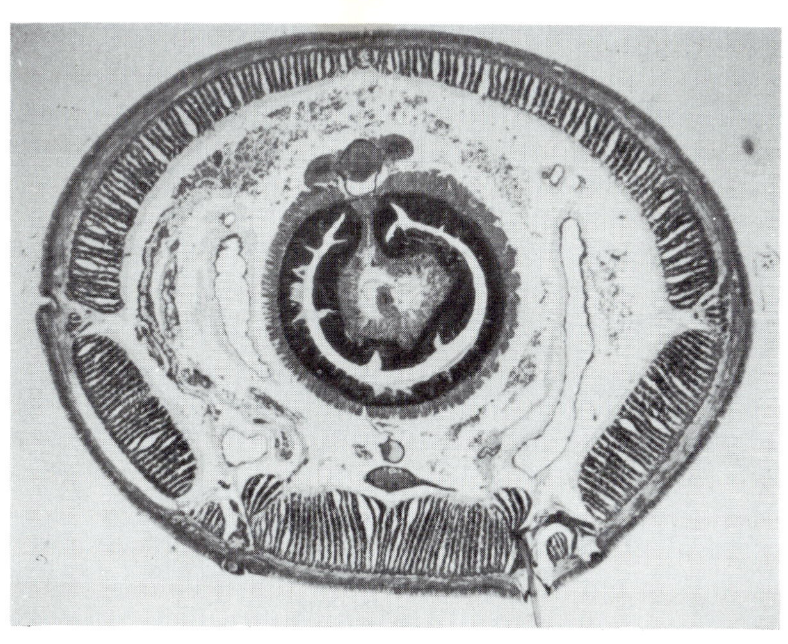

Fig 6.1 TS earthworm intestinal region (x10)

INVESTIGATION 6.1
To identify and interpret features of the internal structure of earthworm

Equipment per student

Prepared slide: TS *Lumbricus* intestinal region

Microscope with light source

Instructions given to students

1 You are provided with a slide of a TS of an earthworm cut through the intestinal region (see also Fig 6.1).

2 Look at it under the microscope. Interpret what you can see in the light of the description of the internal structure given below.

3 *Internal structure of the earthworm.* The body wall is covered by a thin outer cuticle beneath which lies the epidermis made up of one layer of column-shaped cells. Under the epidermis is a layer of circular muscle and below this is a thicker layer of longitudinal muscle divided into nine blocks. Chaetae embedded in pockets or sacs project from the body wall and these can be moved by muscles which pass between the sacs and the body wall.

The alimentary canal (intestine) passes down the centre of the body. In this section a fold called the typhlosole hangs down from the surface of the intestine into its lumen. The intestine is surrounded on the outside by several layers of chloragogenous cells. Internally it is lined with an epithelium. The wall of the intestine also contains muscle layers.

A cavity called the coelom occupies the space between the intestine and body wall. The peritoneum consists of a layer of cells next to the longitudinal muscle of the body wall which lines the coelom. Ventrally in the coelom runs the single median longitudinal nerve cord, the dorsal part of which contains three giant fibres. A dorsal blood vessel is found immediately above the alimentary canal in the layers of chloragogenous cells whilst the ventral blood vessel passes longitudinally immediately below the intestine. Finally, ventral to the nerve cord is a subneural blood vessel.

4 Make a large drawing of the TS earthworm and, using the information given above, label these structures:

dorsal surface
ventral surface
lateral surface
cuticle
epidermis
circular muscle of body wall

typhlosole
epithelium of intestine
chloragogenous cells
coelom
peritoneum
nerve cord

longitudinal muscle of body wall
chaetal muscle
intestinal muscle
lumen of intestine
giant fibres
dorsal blood vessel
ventral blood vessel
subneural blood vessel

Follow-up

1 How many chaetae are there? How are they arranged and in what position? Suggest a function for chaetae.
2 If the longitudinal muscle of the body wall contracts, what effect will it have on overall shape?
3 Chloragogenous cells are full of yellowish-green granules of guanin. With what biological process do you think they are involved?
4 What is the significance of the typhlosole in the intestine?
5 What purpose is served by muscle layers in the intestinal wall?
6 The coelom is full of fluid. Give two possible functions of this fluid-filled cavity.

Assessment

Checklist

1 Correct use of microscope.
2 Production of a large, clear diagram.
3 Accurate labelling on the diagram of the structures listed.
4 Interpretation of the structures identified: evidence of ability to relate structure to function.

This exercise will be most useful for both assessment and training of observational skills if it is given to students who have little or no previous knowledge of earthworm structure. The same technique could also be used with other subject matter, such as the histology of the mammalian alimentary canal, a comparative study of vertebrae, floral anatomy and many other examples. The assessment can be made on the drawn and written record.

Mark scheme

a	Diagram of TS earthworm. Marking will include correct positions and proportions of structures, as well as quality of pencil work.	. . . 5
b	Labels, one mark each.	. . . 20
c	Interpretation:	
	Chaetae: number, arrangement, position, function	. . . 2
	Body wall: longitudinal muscle action	. . . 2
	Chloragogenous cells: function	. . . 1
	Typhlosole: function	. . . 2
	Intestinal wall muscle: function	. . . 1
	Coelomic fluid: function	. . . 2
		35

Other ability which could be assessed

Manipulative skills

Manipulative skills may be assessed by observation during the session to include the way in which the student handles the microscope at high and low magnifications and the techniques used for recording the observed microscopic features on paper.

INVESTIGATION 6.2
A quantitative comparison of the transverse section of a stem and a root

Equipment per student

Prepared slide (A) TS young stem *Ranunculus*

Prepared slide (B) TS young root *Ranunculus*

2 sheets mm graph paper

Microscope with light source

Instructions given to students

1 The prepared slide A is the TS of the **stem** of a young herbaceous dicotyledon.
The prepared slide B is the TS of the **root** of a young herbaceous dicotyledon.

2 On two separate sheets of graph paper make fully labelled diagrams (low power) of both sections to show the distribution of tissues. Do *not* draw individual cells. The two drawings need not be on the same scale, but be very careful to show the proportions of the different tissues as accurately as possible. Include at least half the section.

3 Now count the number of squares of graph paper taken up by the

lignified tissue (or a representative part of it) and by the complete section (or a similar proportion of it) in your diagram. Record your results clearly.

4 For both stem and root calculate the percentage of the cross-sectional area occupied by lignified tissue. Show all the steps in your method of working.

Follow-up

1 Bearing in mind the function of the stem and root, explain the difference in the percentages of the lignified tissue.

2 Try to account for the relative positions of lignified tissue in the stem and root.

Assessment

Checklist

1 Correct use of microscope. Figs 6.2 and 6.3 show transverse sections of stem and root.

2 Preparation of two accurately drawn and well-labelled diagrams.

3 Evidence of the counting of total (or part) cross-sectional area and the proportional area occupied by lignified tissue.

4 Calculations, with working shown, for the percentage of lignified tissue in stem and root.
Results will be of the order of 5% for stem and 1% for root.

5 The percentage results should be explained in terms of the stem remaining upright and needing support against gravity, whereas the location of the root requires less mechanical support.

6 The explanation of the position of lignified tissue should refer to the peripheral columns of support in the stem being able to resist pressures from above, whereas the central core of strength in the root allows a greater resistance to longitudinal stretching caused by water under tension.

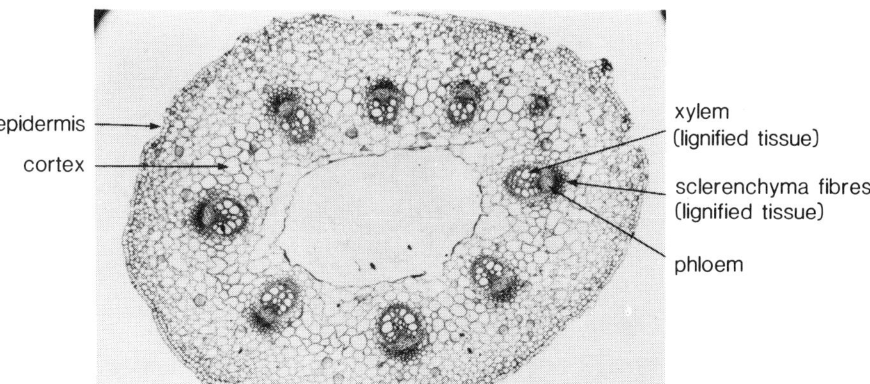

Fig 6.2 TS *Ranunculus* young stem (x60)

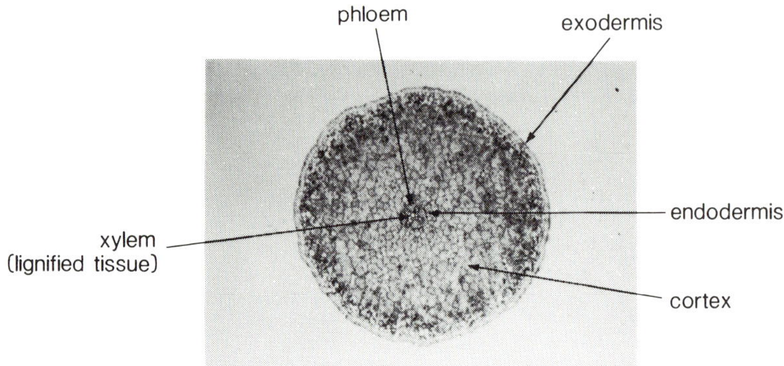

Fig 6.3 TS *Ranunculus* young root (×60)

Mark scheme

a *Stem*

Diagram	. . . 3
Labels, half mark each	. . . 5
Accuracy of proportions	. . . 1
Check the counting and recording of total number of squares	. . . 2
and number of lignified squares	. . . 2

b *Root*

Diagram	. . . 3
Labels, half mark each	. . . 4
Accuracy of proportions	. . . 1
Check the counting and recording of total number of squares	. . . 2
and number of lignified squares	. . . 2
	25

Ideally, for each student the diagrams should be compared carefully with the slides used, but this will depend on time available and the number in the group. Some check should, however, be made in the laboratory on the accuracy and scale of the diagram. The counting technique used and the labelling and quality of the diagram can be marked using the mark scheme.

Other abilities which could be assessed

Manipulative skills
The handling of the microscope, the use of a suitable magnification and the skill of recording the correct aspects of what can be seen on paper can be assessed by observation during the practical session.

Presentation of results and calculations
This is most easily assessed by using a detailed mark scheme with the written work produced. Credit should be given for evidence of accurate counting of the correct areas, method of calculating percentages of lignified tissue and finally the degree to which an acceptable result is obtained for both stem and root.

Interpretation of data
It is advisable that students should consider either average class results or theoretical data at this stage. While it may be valuable for them to try to explain any odd results of their own making, it would be wrong to penalise them twice for errors they have made in drawings and calculations earlier in the exercise.

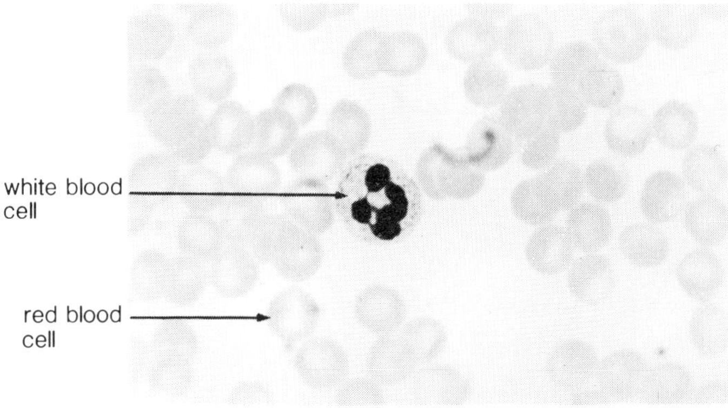

Fig 6.4 Human blood smear (×400)

INVESTIGATION 6.3
The effect of different concentrations of salt solution on red blood cells

Equipment per student

5 cm^3 of each of the following:
1.6% sodium chloride solution, labelled A
0.8% sodium chloride solution, labelled B
0.6% sodium chloride solution, labelled C
0.2% sodium chloride solution, labelled D
Ethanol
Distilled water
Cavity tile
Wax pencil
Teat pipette
Microscope slides and cover slips
Beaker
Capillary tube for collecting blood
Microscope with light source
Sterile lancet
Cotton wool
Filter paper

Instructions given to students

1 Use the pipette to place 3 drops of each of the salt solutions A, B, C and D into each of four cavities on the tile. Label.
2 Rub the side of the first finger with alcohol on cotton wool and allow to dry. Using a sterile lancet, stab the skin and draw blood.
3 Touch the capillary tube onto the blood drop and so draw blood into the tube. Transfer the blood to one of the cavities by blowing down the tube.
4 Repeat until all cavities contain blood. Stir each with the capillary tube starting with D.
5 Rinse the capillary in distilled water by sucking in and out. Blow the water out and touch the end on filter paper to remove excess.
6 With this clean capillary stir the liquid in cavity B. Take up some of this liquid and transfer to a microscope slide. Label, cover with a cover slip and examine under X40 objective.
7 Clean the capillary as before and take further samples in the order A, C and D, stirring each time before sampling. Examine as before.

Follow-up

1 For each solution make a large drawing of two red blood cells to show their relative shapes and sizes. Label.
2 Now examine the blood and salt solutions left in each cavity and, *without further stirring,* describe the appearance of each.
3 Using information from 1 and 2 above, for each solution write a detailed explanation of the results.
4 Which of the four concentrations of salt solution approximates most closely to the concentration of blood plasma? What evidence supports your answer?
5 Outline briefly how you might extend this investigation to determine more accurately the concentration of sodium chloride solution which is isotonic for the blood cells used.

Assessment of Investigation 6.3

Checklist

1 Procedure followed accurately and safely.

2 Microscope used correctly.

3 Diagrams drawn and labelled.

4 Observations explained in terms of the different concentration of salt solutions having different osmotic effects on the red blood cells.

A 0.85% sodium chloride solution is usually given as isotonic with blood plasma.

Students who have not been successful in producing slides for microscopic examination should now be allowed to use those of colleagues or a sample set prepared by laboratory staff. A detailed mark scheme could be applied, which can later be adjusted to the scale required for actual assessment. Figs 6.4 and 6.5 show a blood smear and the effect of sodium chloride solution on red blood cells.

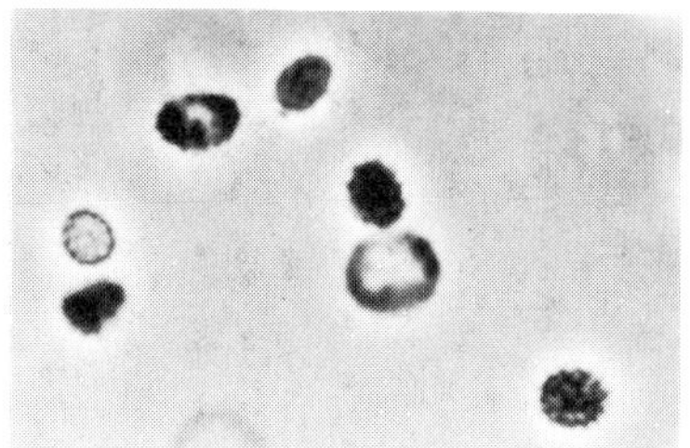

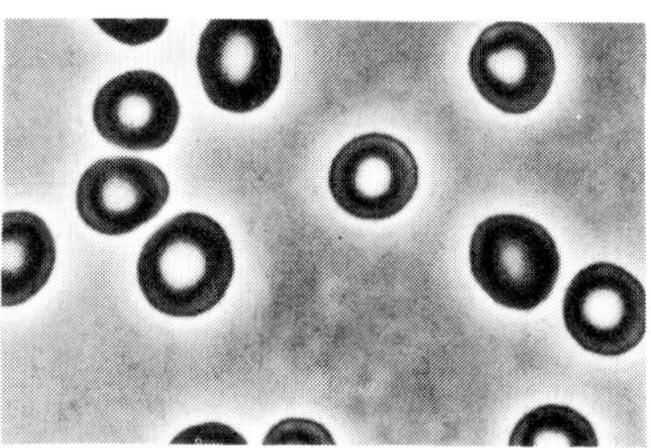

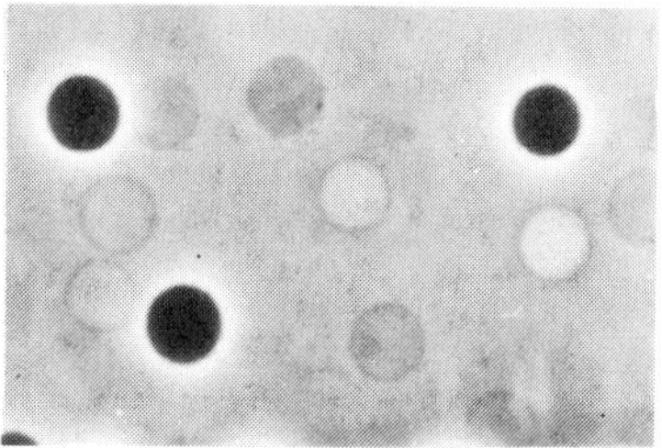

Fig 6.5 Osmotic effects of different salt solutions on red blood cells
(a) Crenation. Effect of a hypertonic solution
(b) Effect of an isotonic solution
(c) Haemolysis. Effect of a hypotonic solution

Mark scheme

a	Drawing and labels for each of A, B, C and D:	(2 + 2) × 4	. . . 16
b	Description of A, B, C and D left in cavity tile:	2 × 4	. . . 8
c	Explanation of results in each case:	4 × 4	. . . 16
			40

Other abilities which could be assessed

Manipulative skills
The handling of the rather small scale apparatus, the taking of the blood sample and the use of the microscope will all provide an opportunity to observe and assess the manipulative skills during the practical period.

Following instructions
Worthwhile results will only be achieved in this investigation if the procedure has been followed methodically. Observation of students at work during the practical and consideration of the final slides will enable a rank order to be worked out.

Experimental design
Once the basic method for this investigation has been undertaken and the data obtained have been discussed, an extension exercise can be tackled. This could be done either as a written account or practically.

Assessment of Investigation 6.4

Checklist

1 Instructions followed carefully. Apparatus and chemicals handled skilfully.

2 Observations recorded fully and with understanding.

3 Observations interpreted. Hypothesis presented in terms of presence of an enzyme in liver and yeast which can be extracted and which catalyses the decomposition of hydrogen peroxide to liberate oxygen.

4 Design of a control experiment involving destruction of proposed enzyme activity by boiling.

If students have followed the worksheet, they will have made accurate observations and recorded and interpreted them. This written work may then be marked and assessed.

INVESTIGATION 6.4
The effect of liver and yeast on hydrogen peroxide

Equipment per student

- 50 cm^3 20 volume hydrogen peroxide
- 1 cm^3 liver
- 1 g dried yeast
- Wooden splint
- 20 cm^3 distilled water
- 1 g clean sand
- 5 test tubes and rack
- Wax pencil
- Bunsen burner
- Forceps
- Filter funnel
- Filter paper
- Pestle and mortar

Instructions given to students

1 Place about 2 cm depth hydrogen peroxide into each of two tubes. Label A and B.

2 Cut the liver into two pieces.

3 To tube A add one of the pieces of liver and to tube B add a pinch of dried yeast.

4 Note any reactions occurring. Insert a glowing splint into each tube, bringing it close to the liquid surface.

   a Record all observations.

   b Put forward a hypothesis which would explain the results so far.

5 Grind together, using a pestle and mortar, the rest of the liver with a little clean sand in 10 cm^3 distilled water. Filter the mixture and collect the filtrate.

6 Place equal volumes of filtrate and hydrogen peroxide in a test tube and again investigate the reaction.

   a Record all observations.

   b Extend or modify your hypothesis to take account of the new results.

7 Put your hypothesis to the test by repeating stages 5 and 6 above, but this time use dried yeast. Design and perform a control experiment.

Follow-up

1 Record details of the control experiment.

2 Record all results and observations.

Mark scheme

a *Hydrogen peroxide and liver or yeast*
Observations: effervescence, relight glowing splint: 2 + 2 . . . 4

Interpretation (hypothesis): biological material bringing about release of oxygen: 2 + 2 . . . 4

b *Filtrate of liver and hydrogen peroxide*
Observations: as before . . . 2

Interpretation: cell extract of liver able to release oxygen, possibility of an enzyme . . . 2

c *Filtrate of yeast and hydrogen peroxide*
Observation: as before . . . 1

Interpretation: cell extract of yeast also liberates oxygen . . . 1

d *Control experiment*
Observation: no release of oxygen . . . 2

Interpretation: cell extract activity destroyed by heating. Therefore enzyme indicated. Present in biological material causing hydrogen peroxide to decompose and liberate oxygen . . . 4

20

Other abilities which could be assessed

Manipulative skills
The ability of students to handle the apparatus, materials and chemicals may be observed and assessed during the practical.

Experimental design
The worksheet leads the student systematically through an open-ended investigation until finally he or she is asked to extend the enquiry further by planning and carrying out a suitable control. The written record completed by the student and handed in for marking will provide the assessment.

Chapter 7

PRESENTATION OF RESULTS AND CALCULATIONS FROM DATA

The student is assessed on his or her ability to select and/or implement the most appropriate method of recording the data collected (or, in some cases, given). In addition, grading is based on the relevance and accuracy of well-explained calculations including elementary statistical analysis.

Presentation of results and calculations from data

The ability to present results in the most sensible and useful way, accurately and with appropriate calculations, is one which students will be learning throughout their course. It will usually be assessed by using a mark scheme with written-up records of a student's practical work. If, however, the student has made errors during the practical or obtained worthless results, he or she should not be penalised further. After correction or discussion (that is, after the assessment situation has turned into a teaching one), the student can be given data from another source such as other students, pooled class data, or results from a book to work on. Results from practical work not done by the class can be treated in the same way to give a written exercise only. Such exercises, followed by discussion, can be useful when a class is in the early stages of a course and learning how to handle and present experimental results.

Students may be told how to present their results, and methods such as tables, graphs, bar charts and annotated diagrams will occur frequently. However, assessment can also be based on the student's ability to select the most appropriate method for particular data.

If calculations are involved, the assessor will be looking for accurate and relevant calculations, with all steps of working shown and explained. In addition, there should be evidence that the student appreciates the errors involved and the limitations of the data processed. As most examining boards allow the use of calculators, students should be able to use them in their calculations.

Practical exercises suitable for the assessment of the presentation of results and calculations from data

Plant physiology

Photosynthesis experiments: carbon dioxide, light intensity and temperature affecting rate

Factors affecting transpiration (or water uptake) using the potometer and other apparatus

Factors affecting seed germination

Stomatal distribution and counts

Effect of temperature on permeability of cell membranes

Gas exchange and energy release during respiration

Water relations using potato discs: change in mass and size

Animal physiology

Determination of lung volumes

Factors affecting respiratory rate of locust

Effect of carbon dioxide and temperature on *Daphnia*

Enzymes

A wide range of exercises is suitable but useful ones include the sensitivity of enzymes to pH and temperature and the effect of substrate concentration on activity

Microbiology

Use of a graticule to calculate area of field of view

Yeast population growth using a haemacytometer

Effect of pH and temperature on yeast growth

Ecology

Factors affecting growth and distribution of organisms such as *Pleurococcus* and *Lemna*

Soil studies: measurement of physical properties

Genetics

Monohybrid and dihybrid crosses using maize cobs, *Sordaria, Drosophila*

Variation studies: use of leaf size, mass and size of seeds

Application of Hardy-Weinberg

INVESTIGATION 7.1
The action of auxin at different concentrations on the growth of roots of cress seedlings

Equipment per student

Sheet of diagrams of cress roots

Length of cotton

Sheet of millimetre graph paper

Instructions given to students

1 Study the diagrams in Fig 7.1. They show cress roots taken from seedlings which have been grown for 24 hours on filter paper soaked in different solutions. Details of each solution are given in Fig 7.1.

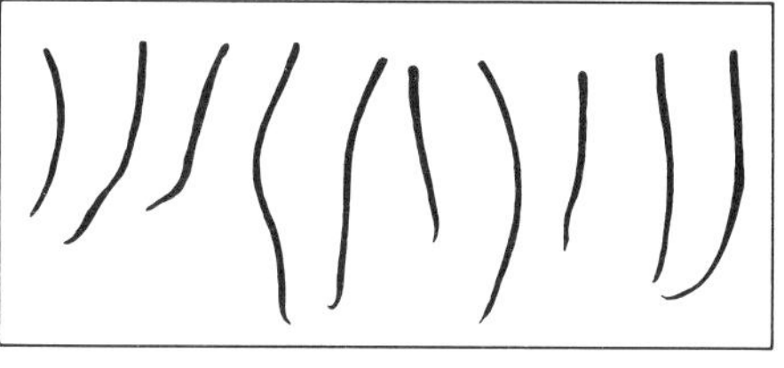

(a)

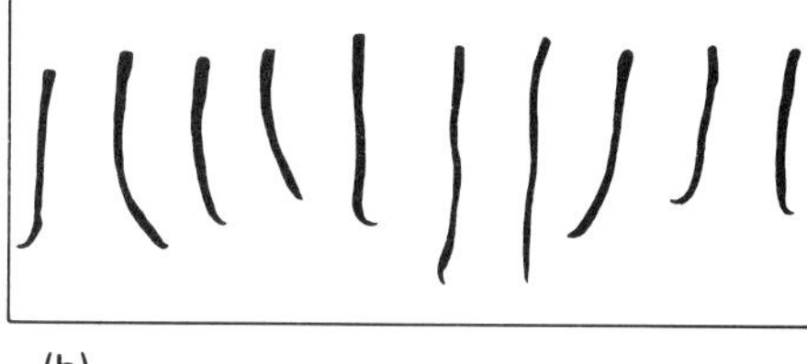

(b)

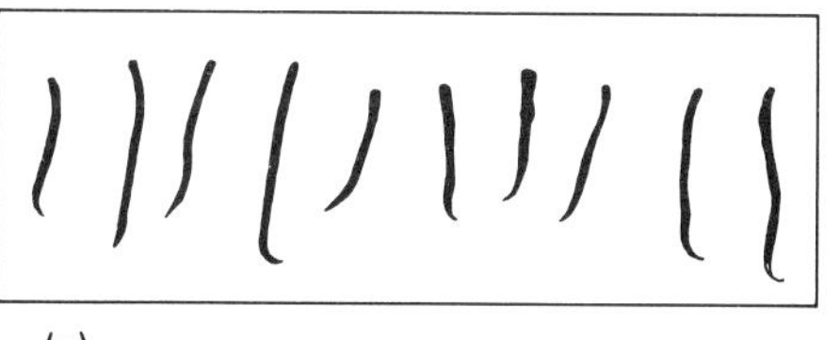

(c)

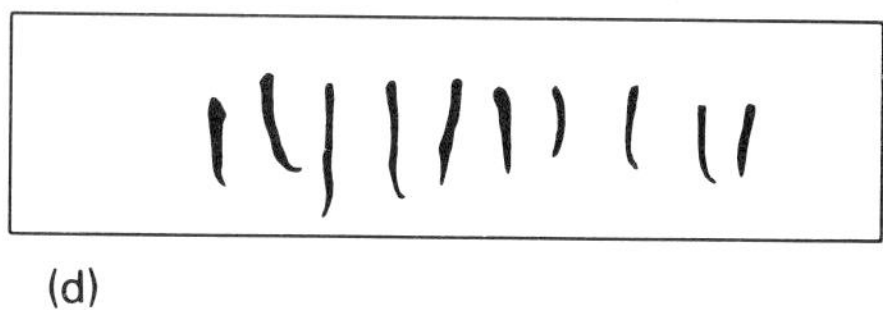

(d)

Fig 7.1 Roots cut from cress seedlings which had been grown for 24 hours on filter paper, soaked in the following solutions of indolyl-3-acetic acid (IAA)
(a) Distilled water (b) 10^{-9} g dm^{-3} IAA
(c) 10^{-7} g dm^{-3} IAA (d) 10^{-5} g dm^{-3} IAA

2 Using the length of cotton and the graph paper, devise a method of measuring, as accurately as possible, the length of each root (in mm).
3 Tabulate your results. For each solution, calculate an average value for root length.
4 Plot a graph of average values for root length against auxin concentration. The values for auxin concentration can be equally spaced along the axis.

Follow-up

1 For each batch of roots there is variation in length. Give *two* reasons why this is so.
2 By reference to the graph, summarise the effect of the auxin treatment on the growth of the cress roots.

Assessment

Checklist

1 Check on method employed to measure root length.
2 Accuracy and care taken in measurement.
3 Tabulation of results: clear and logical.
4 Calculations of averages.
5 Graph construction: scale, axes labelled, best line.
6 Interpretation of information.

Mark scheme

a	Accurate and logical tabulation of results	. . . 5
b	Calculation of average values	. . . 5
c	Construction of graph:	
	Axes correct and labelled	. . . 2
	Suitable scale	. . . 1
	Accurate plotting	. . . 5
	Best curve through points	. . . 2
		20

School experiments to investigate auxin action on plant material can give variable results which may be difficult to analyse. The second-hand data provided here can be useful as a follow-up exercise to such practicals. All students then begin the exercise on an equal footing, regardless of their results in the practical work. Assessment can be done by using a mark scheme on the written material produced.

Other abilities which could be assessed

Manipulative skills
Measurement of root length using cotton and graph paper requires considerable manual skill and patience. These could be tested by checking the accuracy of tabulated results for each student and by observation of students while they are working on the task. From these two sources, grading for this ability should be feasible.

Interpretation of data
Structured questions can also be set on the results of actual practical work. The written responses will indicate the extent to which students can analyse and process their own results.

INVESTIGATION 7.2
The analysis of inhaled and exhaled air using a J-tube gas pipette

Equipment per student

50 cm^3 2M potassium hydroxide
50 cm^3 alkaline pyrogallate solution (under oil)
J-tube gas pipette
Test tube
Length thin flexible tubing
Trough or tank for gas collection
Sheet of millimetre graph paper

Instructions given to students

Analysis of atmospheric air

1 Take the gas pipette (Fig 7.2) and turn the screw three-quarters of the way in.

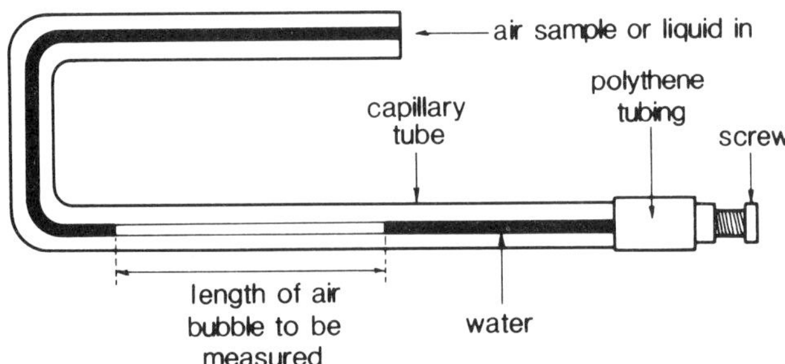

Fig 7.2 J-tube gas pipette

2 Dip the open end of the tube into the container of water and turn the screw outwards so that water is drawn up into the tube. Keep on turning the screw until about 5 cm length of water has been drawn up.
3 To sample the atmospheric air, remove the tube from the water and turn the screw outwards until about 10 cm length of air has been drawn up.
4 Replace the end of the tube in the water and draw up about 5 cm length of water. Adjust the position of the air bubble with the screw so that it is in the long arm of the tube and not in the bend.
5 Measure the length of the bubble on graph paper.
6 Turn the screw inwards so that water is pushed out of the tube until the air bubble almost reaches the open end.
Do not let any air escape.
7 Place the end of the tube in potassium hydroxide solution and draw some up.

8 Keeping the end in the solution, move the air bubble and potassium hydroxide solution backwards and forwards by turning the screw in and out for about 10 minutes.
9 Measure the length of the air bubble again.
10 Remove the potassium hydroxide solution from the tube by turning the screw until the air bubble has almost reached the open end. *Do not let any air escape.*
11 Repeat steps 7 to 10 using alkaline pyrogallate solution instead of potassium hydroxide solution.
12 Wash out the tube several times with water.

NB: POTASSIUM HYDROXIDE SOLUTION AND ALKALINE PYROGALLATE SOLUTION ARE CAUSTIC. TAKE CARE.

Analysis of exhaled air

1 Fill a test tube with water and invert it into a shallow container of water.
2 To collect some exhaled air, position the piece of tubing as in Fig 7.3 and then blow through it until all the water in the test tube has been replaced by air.

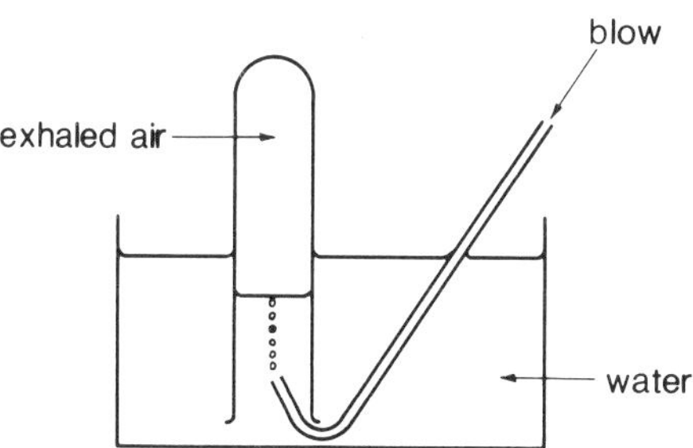

Fig 7.3 Collection of exhaled air

3 Repeat steps 1 and 2 as for atmospheric air.
4 To draw up exhaled air into the gas pipette, place the open end inside the test tube as shown in Fig 7.4 overleaf and draw up about 10 cm length of air.
5 Repeat steps 4 to 12 as for atmospheric air.

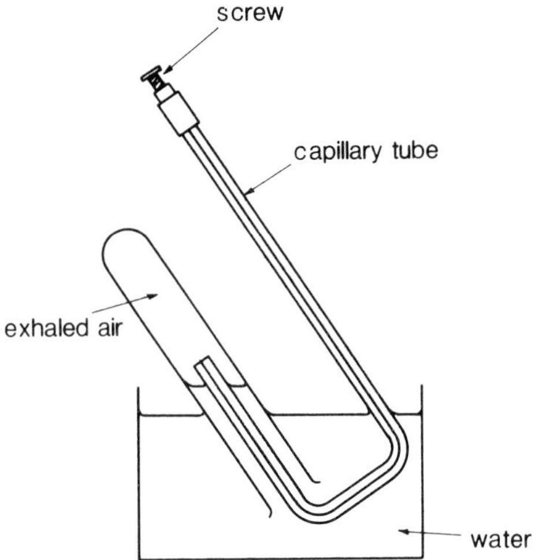

Fig 7.4 Exhaled air is drawn into the gas pipette

Follow-up

1 What is the purpose of using potassium hydroxide solution in this experiment?
2 What is the purpose of using alkaline pyrogallate solution?
3 Why is it important to use the potassium hydroxide solution first?
4 For atmospheric air:
   a record the initial length of the air bubble.
   b record the length after potassium hydroxide treatment.
   c record the length after alkaline pyrogallate treatment.
   d calculate the percentage of carbon dioxide present in the air sample using this technique.
   e calculate the percentage of oxygen present in the same sample.
5 For exhaled air: repeat the recordings and calculations as for atmospheric air above.
6 List five possible sources of error in this practical procedure. For each suggest a modification to the technique which could minimise the error.

Assessment

Checklist

1 Methodical approach to following instructions.

2 Safe handling of equipment, and particularly of chemicals. Tidy bench work.

3 Presentation of results in a clear table.

4 Calculations performed with full explanation of steps taken.

5 Appreciation of limitations of the procedure and plans for modification of technique.

Not all pupils will have satisfactorily completed the practical in which case second-hand data could be given to them, or an average of class results used. A similar time-allowance should be allocated to each pupil for the written work.

Mark scheme

a	Recording and tabulation of results:		
	Atmospheric air		. . . 2
	Exhaled air		. . . 2
b	Calculations: each step explained and carried out correctly		
	Atmospheric air:	% carbon dioxide	. . . 3
		% oxygen	. . . 3
	Exhaled air:	% carbon dioxide	. . . 3
		% oxygen	. . . 3
c	Final result: some reference made to limitations in terms of validity of % calculated		
	Atmospheric air:	% carbon dioxide	. . . 1
		% oxygen	. . . 1
	Exhaled air:	% carbon dioxide	. . . 1
		% oxygen	. . . 1
			20

Other abilities which could be assessed

Manipulative skills
This is a very good exercise for testing manual dexterity, although students would need some time beforehand to experiment with the apparatus if they are to complete the entire practical in the usual time allocated. The manipulation of the pipette and air bubble, handling of the potentially harmful chemicals and the measuring procedure all provide an opportunity for assessment.

Following instructions
This is also an excellent practical for putting pupils in a situation where they are asked to work through a series of operations, although this ability should not be assessed at the same time as manipulative skill.

Experimental design
It is hoped that students will see both the limitations of the experiment itself and their execution of it. Section 6 of the follow-up enables them to display this knowledge and provides an opportunity to suggest alternative procedures.

INVESTIGATION 7.3
The separation of leaf pigments by chromatography

Equipment per student

10 cm^3 acetone
5 cm^3 leaf pigment solvent: (1 part 90% acetone to 9 parts petroleum ether)
Fresh green leaves (geranium, nettle)
Beaker
Bunsen burner and tripod
Forceps
Scissors
Small pestle and mortar
Filter funnel
Conical flask
Boiling tube with cork (fitted with pin)
Boiling tube rack
Filter paper
Chromatography paper.

Instructions given to students

1 Take 5 large fresh green leaves, eg geranium or nettle. Dip them into boiling water for 2 minutes.
2 Chop up the lamina of each leaf into small pieces. Grind these pieces, using a pestle and mortar, with a minimum volume of acetone to form a very concentrated extract.
3 Filter and collect the filtrate.

4 Cut a strip of chromatography paper of sufficient length to reach almost to the bottom of the boiling tube and of such width that the edges do not touch the sides.
5 Draw a pencil line across the strip 3 cm from one end. Fold the other end over through 90° and by means of a pin attach it to a cork as in Fig 7.5.

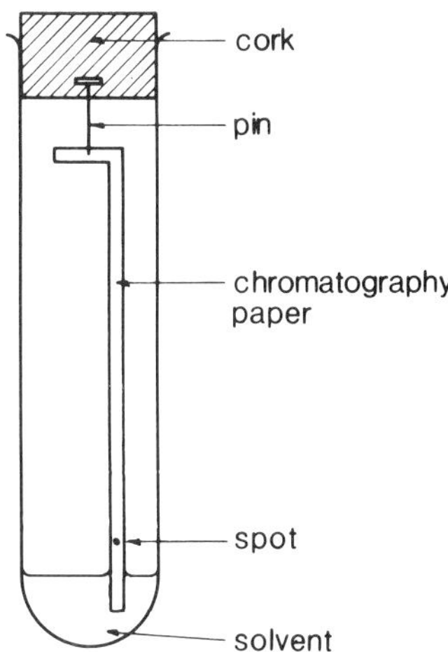

Fig 7.5 Section through the boiling tube and contents for leaf chromatography

6 Remove the paper. Using the head of a pin as a dropper, place a drop of 'leaf' solution at the centre of the pencil line. Let the drop dry, then apply a second on top of the first. Repeat this for about 10 minutes, building up a small but concentrated area. This is most important.
7 Pour some solvent (acetone and petroleum ether) into the boiling tube to a depth of about 1.5 cm. Seal the tube with the cork and leave to provide a saturated atmosphere inside.
8 Suspend the strip of paper in the boiling tube. The bottom edge of the paper should dip in the solvent, but the coloured spot should not touch it.
9 Leave for about 10 minutes. When the solvent is about 2 cm from the top, remove the strip and rule a pencil line to mark the final solvent front. Allow the paper to dry.
10 The pigments should separate in the order:
chlorophyll b (green)
chlorophyll a (blue/green)
xanthophyll (yellow/brown)
carotene (yellow)
11 Lightly pencil in the leading edge and the outline of each pigment. Label.

12 For each, calculate the Rf value:

$$Rf = \frac{\text{Distance moved by substance from original position}}{\text{Distance moved by solvent from same position}}$$

Follow-up

Suggest a follow-on method which you might use to investigate the presence of any additional pigments in the leaf.

Assessment

Checklist

1 Efficient and safe handling of apparatus and chemicals.

2 Procedure completed according to instructions.

3 Chromatogram prepared, outlines of pigments pencilled in and labelled.

4 Measurements made, calculations made and Rf values recorded.

chlorophyll b (green)	Rf = 0.45
chlorophyll a (blue/green)	Rf = 0.65
xanthophyll (yellow/brown)	Rf = 0.71
carotene (yellow)	Rf = 0.95

The clarity of the chromatograms produced will depend on the accuracy of the techniques employed. Assessment of the actual results and calculations made must be based on the chromatogram of each student and not on theoretical results. If the chromatograms are pencilled in and labelled clearly as instructed, they can be marked at a later stage.

Mark scheme

a	Clarity and accuracy of chromatograms		. . . 4
b	Interpretation and accurate labelling		. . . 4
c	For each pigment:		
	measurement of pigment front	4×1	. . . 4
	measurement of solvent front		. . . 1
d	For each pigment:		
	correct application of formula	4×1	. . . 4
	accurate calculations	4×1	. . . 4
e	Accuracy of Rf values calculated compared with theoretical	4×1	. . . 4
			25

Other abilities which could be assessed

Manipulative skills
The production of the leaf extract, preparation of the chromatography paper and assembling of the apparatus and chemicals test the manual dexterity of the student very fully. A mark could be given during the session, although the standard of the final chromatogram will also indicate the effectiveness of the techniques employed.
Following instructions
Competence in following instructions will also be evident from close observation of the students at work and again by analysis of the chromatograms produced. It would be unwise to assess this ability in addition to manipulative skills, and the choice depends on the needs of the assessor at the time.

INVESTIGATION 7.4
To determine the osmotic potential of cell sap in onion epidermal cells

Equipment per student

Distilled water
10 cm^3 of each of the following sucrose solutions: 0.3 M, 0.35 M, 0.4 M, 0.5 M, 1.0 M
Fleshy scale leaf from onion
5 stoppered tubes
Watch glass
Razor blade
Wax pencil
Forceps
Microscope slides
Microscope with light source

Instructions given to students

1 From the onion take a fleshy scale leaf and with the razor blade cut the inner epidermis into six squares, each of side about 5 mm.
2 Remove these squares carefully with forceps and place them in a watch glass of distilled water.
3 Label five specimen tubes as follows: 0.3 M, 0.35 M, 0.4 M, 0.5 M and 1.0 M. Place about 10 cm^3 of the appropriate sucrose solution into each.
4 Add one square of the epidermis to each tube, put in the stopper and shake gently. Leave for approximately 30 minutes.

5 After this period, remove the square of epidermis from the 0.3 M sucrose solution and mount it in a drop of the same solution on a microscope slide. Label.

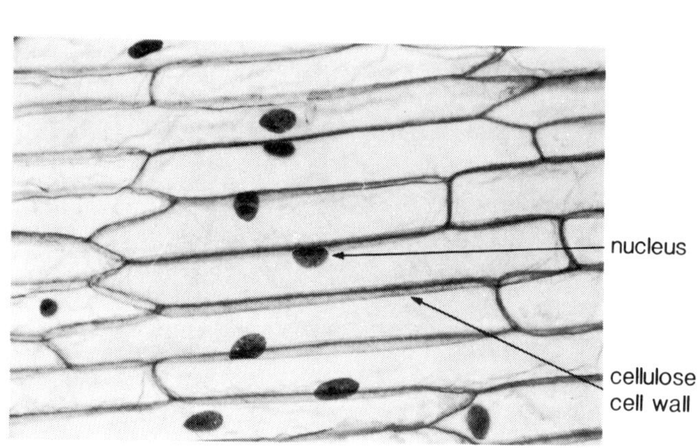

Fig 7.6 *Allium* bulb scale epidermis (×250)

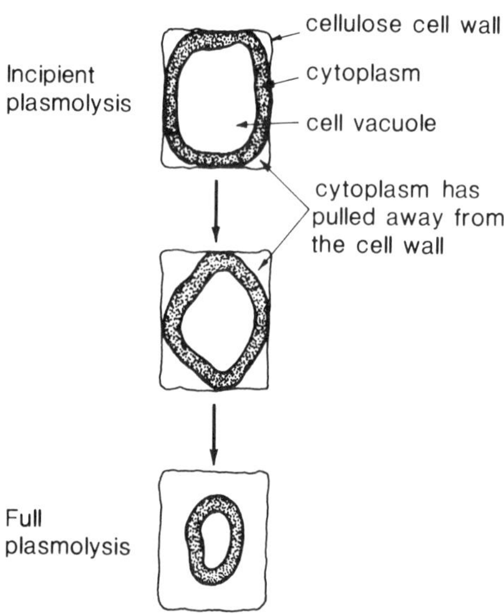

Fig 7.7 Stages in plasmolysis

6 Observe under low power magnification. Fig 7.6 shows these cells under high magnification.
   a Count and record all cells present in the field of view.
   b Count and record the number of those cells which show any visible signs of plasmolysis, however small (see Fig 7.7).
7 Repeat this procedure for each of the remaining four tubes in turn.

Follow-up

1 For each solution calculate the percentage of cells plasmolysed.
2 Plot a graph of percentage of cells plasmolysed against molarity of sucrose solution.
3 For practical purposes 50% plasmolysis of the tissue may be regarded as the point of incipient plasmolysis. The solution causing this is considered to have the same osmotic potential as the cell sap of the tissue. From the graph read off the molarity of sucrose which corresponds to 50% plasmolysis.
4 Why is it important to calculate percentage plasmolysis for each solution?
5 Why should the squares of epidermis be placed in distilled water before being put into experimental solutions?
6 How would you proceed by further experiment to obtain a more precise value for osmotic potential?
7 Why could the same procedure not be used to measure the osmotic potential of a protozoan such as *Amoeba?* Suggest a method which might be used.

Assessment

Checklist

1 Procedure followed carefully and exercise completed satisfactorily.
2 Microscope used correctly.
3 Cell counts made. Results logically tabulated.
4 Calculations performed to give percentages.
5 Graph plotted with precision. Axes correct and labelled.
6 Reading taken from graph.
7 Evidence of an understanding of procedure employed.
8 Extension and modification of technique used.

The theoretical background to this procedure will need to be explained at some stage, and probably clarification and guidance will have to be given on the definition of plasmolysis of a cell during the practical. There should then be little problem over the counts being made and tabulated, calculations carried out and the graph plotted. The written record can be marked and assessed at a later stage.

Mark scheme

For each sucrose solution

a	Table of results:	
	Total cells in field of view	5×1 . . . 5
	Number of cells plasmolysed	5×1 . . . 5
b	Calculation of percentages in each case	5×1 . . . 5
c	Construction of graph:	
	Axes correct, labelled, suitable scale	. . . 2
	Points plotted accurately and clearly	. . . 4
	Best line through	. . . 2
d	Reading of molarity at 50% plasmolysis	. . . 2
		25

Other abilities which could be assessed

Manipulative skills

The preparation of the strips of inner epidermis, their transfer into and out of solutions, together with careful and efficient microscope manipulation will test manipulative skills fully. Marks could be awarded during the session.

Following instructions

Although it is not advisable to assess this simultaneously with manipulative skills, a rank order of students could be achieved by observation during the procedure.

Experimental design

The student is asked to suggest an extension to this investigation to obtain more precise values, and also an alternative procedure for work with animals. The written responses will indicate planning ability.

Chapter 8

INTERPRETATION OF DATA

Under this heading, pupils are asked to analyse experimental results of both a qualitative and quantitative nature and to draw significant conclusions. Assessment will discriminate between those who appreciate the limitations of the data and the weaker candidates who, regardless of errors inherent in the procedure, make bold statements of fact based on theory work.

Interpretation of data

The information collected as a result of completing a practical procedure may be descriptive, in the form of a diagram or, more frequently, as numerical data. It is the student's ability to draw meaningful conclusions from such information which is assessed here, and there is no shortage of suitable data to work on.

Although the teacher may well question the student during the course of the practical about the findings, it is the written record which will provide most of the assessment evidence. Guidance should be given to the class on how to interpret, analyse and draw conclusions as, if left to themselves, only the best students will have ideas on how to tackle the interpretation. The exercise will be much more productive if the teacher asks a series of structured questions which stimulate the student to analyse the results more logically and critically. Undoubtedly, a structured approach aids teacher marking too. It also enables one to move from simple to more complex points, giving an arrangement of questions which allows the teacher to detect, by the cut-off point, how far an individual has got in his or her thinking.

Carefully written questions, designed to test different thinking skills, can be attached to class results to give a data interpretation problem of the teacher's own devising. Duplicated and filed, a bank of such problems can be useful to draw on when time is too short to do the experiment or for some other reason results are not available, as experience for absentees or extra work for particular students, or as a written class homework or examination exercise.

When students set about interpreting their own results, they must be aware of sources of error, drawbacks of the methodology, the constraints of the situation and so on, depending on the particular investigation. The more able students will appreciate this, giving a factor which may often be used in producing a rank order.

Practical exercises suitable for the assessment of interpretation of data

Any investigation which involves the production of descriptive, visual or numerical data will be suitable for assessment, and many examples are given in previous chapters. The more useful practical areas are summarised below, and sources of second-hand data can be found in Appendix 3.

Genetics

Analysis of monohybrid and dihybrid crosses using a variety of organisms

Problems involving quantitative studies of variation such as PTC tasting, tongue rolling, mass of seeds, leaf size

Plant physiology

Factors affecting photosynthesis rate

Conditions necessary for germination

Transpiration and water uptake investigations

Water culture experiments

Investigations into the pathways taken by water, mineral salts and organic food through the plant

Plant hormone experiments. Regions of detection of stimuli and regions of response. Use of coleoptiles and auxin application.

Water relations

Water potential investigations

Plasmolysis

Haemolysis

Respiration experiments

Gas analysis

Tidal volume

Determination of metabolic rates

RQ and Q_{10} values

Vital capacity

Anaerobic respiration in yeast

Ecology

Distribution studies

Feeding relationships

Competition

Pollution studies

Seashore zonation

In addition to the pooling of class results, second-hand data (data which have not been collected by the class) are very useful. Sources of such data include published books of problems and questions from past examination papers. Whilst it is probably satisfying for the student if the data are related to procedures already carried out, this is by no means essential and an approach which does not first involve the students in practical work may be used from time to time for assessment purposes.

INVESTIGATION 8.1
The identification of an enzyme using a dried skimmed milk–agar substrate

Equipment per student

2 cm^3 enzyme suspension (4% trypsin solution freshly made)
M/10 sodium carbonate solution
M hydrochloric acid
Distilled water
Agar powder
Fresh dried skimmed milk
Balance
Paper punch
Stop-clock
Spatula
Bunsen burner
Tripod
Beaker
50 cm^3 measuring cylinder
Stirring rod
Petri dish and lid
2 cm^3 syringe or pipette
Watch glass
Forceps
Teat pipette
Universal indicator paper
Filter paper

Instructions given to students

The object of this experiment is to identify a given enzyme by its action on an agar substrate of dried skimmed milk. The white colour is due to casein.

1 Preparation of substrate: weigh out 0.5 g dried skimmed milk and 0.25 g agar powder. Add these to 25 cm^3 water in a beaker and stir. Boil for 2–3 minutes. This amount is sufficient for up to 3 plates.
2 Pour a *thin* layer of jelly substrate into a clean Petri dish. Allow to cool and set.
3 Using a syringe, place 2 cm^3 enzyme suspension in a watch glass. Test the pH with universal indicator paper and adjust to pH 7–8 with M/10 sodium carbonate solution. Record pH.

4 Take 3 small circles of filter paper and label A, B and C.
5 Using forceps, dip A into the enzyme and place lightly on the jelly substrate.
6 Add 1 M hydrochloric acid to the enzyme in the watch glass until pH 1–3 is achieved. Test with universal indicator paper and record.
7 Dip B into the acidified enzyme and place on the substrate.
8 Repeat the procedure using C dipped into distilled water.
9 Cover the Petri dish with a lid and leave in a warm place for 30 minutes.

Follow-up

1 Record your results by means of a labelled diagram. Then consider the pooled results of the class and attempt to explain them.
2 The enzyme used is to be found in the human alimentary canal. Identify the enzyme as far as possible, giving your reasons.
3 Explain the purpose of filter paper C.
4 Suggest, with reasons, a more suitable alternative to filter paper C.

Assessment

Checklist

1 Logical sequence of operations.
2 Careful handling of apparatus and material.
3 Preparation of a well-drawn, fully labelled record of Petri dish and contents (Fig 8.1).

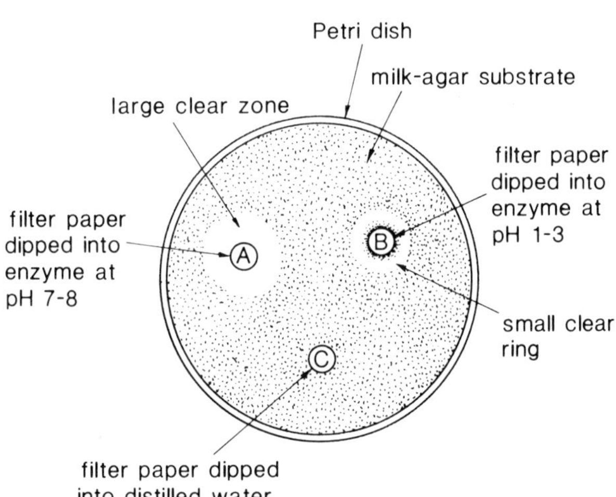

Fig 8.1 Results of enzyme action on the milk–agar substrate

4 Data to be explained in terms of the protein casein digested by the enzyme producing a clear zone in the agar. The proteinase favours neutral to alkaline conditions and is probably located in the human small intestine.

It is important that, in assessing this ability, all students should start on an equal footing. Therefore a consideration of results obtained in the class as a whole should be made. A specific mark scheme is then developed for the step by step deduction made in identifying the enzyme. Students should be penalised for making statements which are not substantiated by their data.

Mark scheme

a	Explanation of observations for each of areas around discs A, B and C	3 × 3	. . . 9
b	Logical series of deductions to identify the enzyme as far as possible		. . . 6
c	Explanation of the purpose of disc C		. . . 5
			20

Other abilities which could be assessed

Observation and recording
This ability could be assessed by comparing the results in each agar plate with the corresponding diagram.

Following instructions
This could be assessed by observing the students' work throughout the practical. Notes on progress, taking into account any help which was required, should enable a rank order to be produced.

Experimental design
Section 4 in the follow-up provides students with an opportunity to suggest modifications to the procedure. As in Investigation 7.2, this type of approach provides a neat and simple way of introducing the idea of experimental design.

INVESTIGATION 8.2
The action of saliva on starch solution in a 'model gut'

Equipment per student

10 cm^3 1% starch 'solution' (containing 0.1% sodium chloride solution)
Benedict's solution
Iodine in potassium iodide solution
Distilled water
25 cm length narrow Visking tubing
10 cm^3 measuring cylinder
Teat pipette
Cotton
Paper clips
Rubber band
Waterbath at 37 °C
Wax pencil
2 test tubes

Instructions given to students

1 Prepare a dilute solution of salivary amylase: rinse your mouth out with water and then chew for a few minutes on a clean rubber band. Now place two 'dribbles' of saliva in the measuring cylinder and dilute with 10 cm^3 distilled water.

2 Take the length of Visking tubing and tie a very tight knot in one end. Using the pipette, fill the tubing to three-quarters of its capacity with a 1% starch solution. Mark this level with a piece of cotton tied loosely to the outside of the tubing (see Fig 8.2).

3 Top up the tubing with saliva solution. Take care *not* to mix the two solutions. Fasten the open end with a paper clip and rinse the outside of the tubing thoroughly with water.

4 Place the Visking tubing and contents in a test tube of distilled water, making sure that the open end is above the level of the water. Incubate at 37 °C in a water bath for at least 30 minutes.

5 Repeat the procedure, but mix the starch and saliva solutions thoroughly in the Visking tubing before placing in the water bath.

6 When the two test tubes have been left for the same period of time, remove a sample of water from each test tube and test each separately with Benedict's solution and with iodine solution. Record your results.

7 Finally, add iodine solution to the water around the Visking tubing and leave for 10 minutes. Record your results by means of a labelled diagram of the apparatus.

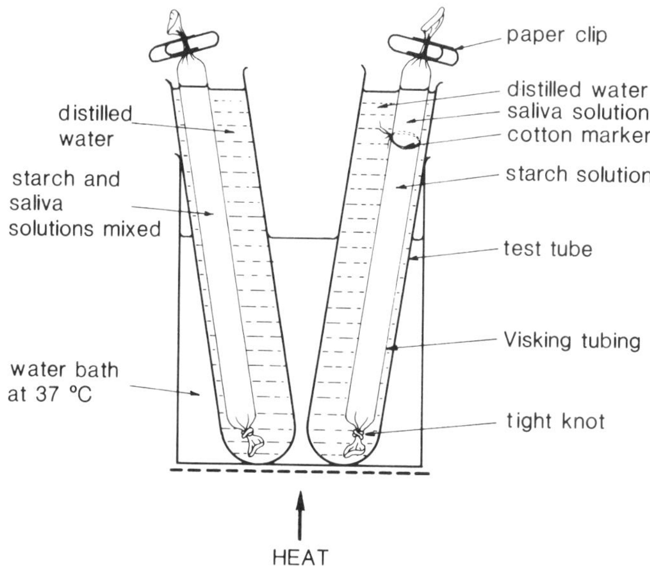

Fig 8.2 'Model gut' apparatus

Follow-up

1 For part 6 compare the results for the two parts of the experiment and explain them fully.
2 For part 7, similarly compare labelled diagrams and explain the results obtained.
3 Summarise in one or two sentences the deductions you can make about the experiment as a whole.

Assessment

Checklist

1 Precise handling of materials and chemicals.
2 Procedure carried out accurately and with care.
3 Tests carried out correctly. Results presented according to worksheet.
4 Data interpreted: salivary amylase digestion of starch when the two solutions are in maximum contact at an optimum temperature, to produce reducing sugar which diffuses through the differentially permeable membrane into the surrounding medium.

Some students will have made errors during the practical procedure or may not have completed it in the time available. In assessing interpretation, a standard set of results or class results as a whole should be used. The written explanation of the data may then be assessed by reference to a mark scheme.

Mark scheme

a	*For unmixed solutions*			
	Conclusions drawn from:	Benedict's test	. . .	1
		Iodine test	. . .	1
	For mixed solutions			
	Conclusions drawn from:	Benedict's test	. . .	1
		Iodine test	. . .	1
b	*For unmixed solutions*	Diagram with labels	. . .	3
		Inferences	. . .	2
	For mixed solutions	Diagram with labels	. . .	3
		Inferences	. . .	2
c	Summary interpretation of results		. . .	6
				20

Other abilities which could be assessed

Manipulative skills
The filling of the Visking tubing and assembling of the apparatus require considerable manual dexterity which could be assessed during the practical session.

Following instructions
The experimental method must be followed methodically and thoughtfully if the exercise is to have any purpose at all. Observations of the students will quickly reveal those who have made errors or who need assistance, and they can be graded accordingly. It would be difficult and unfair to assess this skill simultaneously with manual dexterity.

INVESTIGATION 8.3
The analysis and interpretation of results from a genetics exercise involving *Drosophila*

Equipment per student

Each student will require the worksheet and access to chi-squared tables.

Instructions given to students

1 A series of crosses using the fruit fly *Drosophila* were carried out.
   a Parental generation: a cross was made between pure-breeding wild type males (long wing and red eye) and pure-breeding mutant virgin females (vestigial wing and white eye) (Figs 8.3 and 8.4). The F_1 generation was scored as the flies emerged.
   b The F_1 flies were then mated to produce the F_2 generation.
   c The types and numbers of *Drosophila* produced in the F_1 and F_2 generations were as shown in Table 8.1 overleaf.

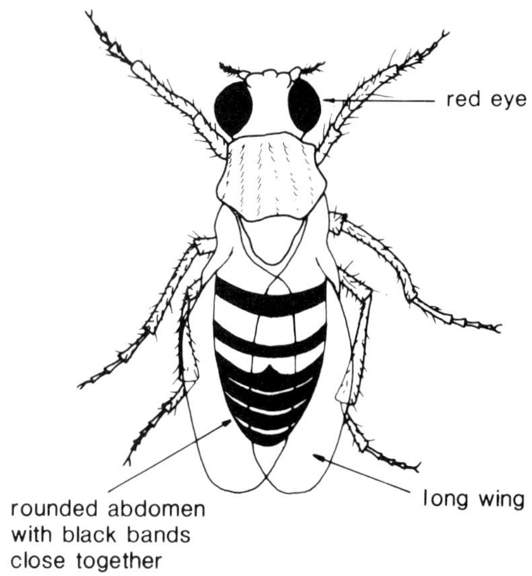

Fig 8.3 Male *Drosophila*: long wing and red eye

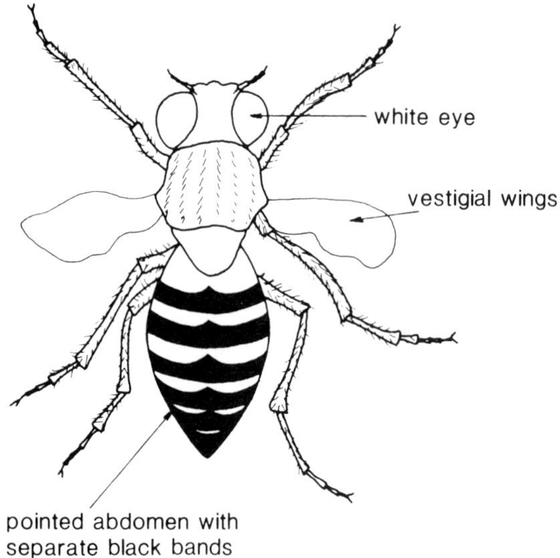

Fig 8.4 Female *Drosophila*: vestigial wing and white eye

Table 8.1 Results from a genetics exercise using *Drosophila*

		Long wing Red eye	Vestigial wing Red eye	Long wing White eye	Vestigial wing White eye
F_1	Male			1491	
	Female	1532			
F_2	Male	594	178	608	194
	Female	550	181	632	209
	Total				

Follow-up

1 *a* For F_1 and F_2 generations together, calculate the male to female ratio.

b Assuming that the mechanism for sex inheritance in *Drosophila* is similar to that for humans, explain the ratio.

2 *a* Using evidence from the F_1 and F_2 generations, what assumption might you make about the alleles for wing length?

b Considering the inheritance of wing length alone, explain the crosses made which produced F_1 and F_2 generations.

c Would the ratios have been the same if the parental flies had been vestigial wing males and long wing females? Explain.

3 *a* In the F_1 generation, calculate the red eye to white eye ratio. Is there any other relevant observation to make?

b A similar cross used vestigial wing, white eye males and long wing, red eye females as parents. All the offspring in the F_1 generation, both male and female, were long wing and red eye.

c Using evidence 3*a* and 3*b*, what assumption can you make about the inheritance of eye colour? Considering the inheritance of eye colour alone, explain the crosses made which produced the F_1 and F_2 generations.

4 Finally, considering the inheritance of wing length and eye colour together, explain the crosses made and give the theoretical ratios for the F_1 and F_2 generations.

5 *a* Using the chi-squared test, investigate the deviation of the observed results in the F_2 (total male and female together) from the expected results as proposed by the theoretical ratios for F_2. The null hypothesis will state that there is no difference between the observed and expected data.

$$\chi^2 = \sum\left[\frac{(\text{observed} - \text{expected})^2}{\text{expected}}\right]$$

Calculate the value of χ^2.

b Consult a significance table for χ^2 (Table 8.2 overleaf). In this case there are 4 categories and so there will be 3 degrees of freedom and, within the limits of biological variability, it is usual to accept 5% (0.05) as a significant probability level.

What is the value of χ^2 given in the table for 3 degrees of freedom and at the 5% significance level? Record this value.

c If the value of χ^2 calculated is greater than the value in the table, the null hypothesis must be rejected. From your calculations, do you accept or reject the null hypothesis?

d Is there then a significant difference at the 5% level between observed and expected results?

e Consequently, do the original assumptions regarding the inheritance of eye colour and wing length stand?

6 Suggest 3 reasons why you would expect some differences between practical results and theoretical ratios.

Table 8.2 Distribution of χ^2

Degrees of freedom	Probability, *p*											
	0·99	0·98	0·95	0·90	0·80	0·50	0·20	0·10	0·05	0·02	0·01	0·001
1	0·000	0·001	0·004	0·016	0·064	0·455	1·64	2·71	3·84	5·41	6·64	10·83
2	0·020	0·040	0·103	0·211	0·446	1·386	3·22	4·61	5·99	7·82	9·21	13·82
3	0·115	0·185	0·352	0·584	1·005	2·366	4·64	6·25	7·82	9·84	11·35	16·27
4	0·297	0·429	0·711	1·064	1·649	3·357	5·99	7·78	9·49	11·67	13·28	18·47
5	0·554	0·752	1·145	1·610	2·343	4·351	7·29	9·24	11·07	13·39	15·09	20·52
6	0·872	1·134	1·635	2·204	3·070	5·35	8·56	10·65	12·59	15·03	16·81	22·46
7	1·239	1·564	2·167	2·833	3·822	6·35	9·80	12·02	14·07	16·62	18·48	24·32
8	1·646	2·032	2·733	3·490	4·594	7·34	11·03	13·36	15·51	18·17	20·09	26·13
9	2·088	2·532	3·325	4·168	5·380	8·34	12·24	14·68	16·92	19·68	21·67	27·88
10	2·558	3·059	3·940	4·865	6·179	9·34	13·44	15·99	18·31	21·16	23·21	29·59
11	3·05	3·61	4·58	5·58	6·99	10·34	14·63	17·28	19·68	22·62	24·73	31·26
12	3·57	4·18	5·23	6·30	7·81	11·34	15·81	18·55	21·03	24·05	26·22	32·91
13	4·11	4·77	5·89	7·04	8·63	12·34	16·99	19·81	22·36	25·47	27·69	34·53
14	4·66	5·37	6·57	7·79	9·47	13·34	18·15	21·06	23·69	26·87	29·14	36·12
15	5·23	5·99	7·26	8·55	10·31	14·34	19·31	22·31	25·00	28·26	30·58	37·70
16	5·81	6·61	7·96	9·31	11·15	15·34	20·47	23·54	26·30	29·63	32·00	39·25
17	6·41	7·26	8·67	10·09	12·00	16·34	21·62	24·77	27·59	31·00	33·41	40·79
18	7·02	7·91	9·39	10·87	12·86	17·34	22·76	25·99	28·87	32·35	34·81	42·31
19	7·63	8·57	10·12	11·65	13·72	18·34	23·90	27·20	30·14	33·69	36·19	43·82
20	8·26	9·24	10·85	12·44	14·58	19·34	25·04	28·41	31·41	35·02	37·57	45·32
21	8·90	9·92	11·59	13·24	15·45	20·34	26·17	29·62	32·67	36·34	38·93	46·80
22	9·54	10·60	12·34	14·04	16·31	21·34	27·30	30·81	33·92	37·66	40·29	48·27
23	10·20	11·29	13·09	14·85	17·19	22·34	28·43	32·01	35·17	38·97	41·64	49·73
24	10·86	11·99	13·85	15·66	18·06	23·34	29·55	33·20	36·42	40·27	42·98	51·18
25	11·52	12·70	14·61	16·47	18·94	24·34	30·68	34·38	37·65	41·57	44·31	52·62
26	12·20	13·41	15·38	17·29	19·82	25·34	31·80	35·56	38·89	42·86	45·64	54·05
27	12·88	14·13	16·15	18·11	20·70	26·34	32·91	36·74	40·11	44·14	46·96	55·48
28	13·57	14·85	16·93	18·94	21·59	27·34	34·03	37·92	41·34	45·42	48·28	56·89
29	14·26	15·57	17·71	19·77	22·48	28·34	35·14	39·09	42·56	46·69	49·59	58·30
30	14·95	16·31	18·49	20·60	23·36	29·34	36·25	40·26	43·77	47·96	50·89	59·70

Assessment

Checklist

1. Calculation of male/female 50 : 50 ratio and explanation of sex inheritance in terms of autosomes and X and Y chromosomes.
2. Deduction that long wing allele dominant to vestigal allele and explanation of the crosses producing F_1 and F_2.
3. Deduction that eye colour is a sex-linked factor and genetical explanation of F_1 and F_2.
4. Explanation of the inheritance of wing length and eye colour together producing a 3:1:3:1 ratio in F_2 if all assumptions so far are correct.
5. Statistical analysis of observed results against theoretical ratio. Calculated value of χ^2 (7.38) not greater than 7.82 (figure for 5% significance level and 3 degrees of freedom). The null hypothesis stands. No significant difference

between observed and expected results. Statistical analysis does not contradict the original genetical assumptions.

6 Appreciation of the limitations of the experiment and techniques used.

Ideally, this exercise will be given to students who are familiar with the theory of sex inheritance and dihybrid crosses. In addition it is helpful if they have had experience of practical procedures involving *Drosophila*. In fact, they could have set up or even completed this particular exercise for themselves. If so, the table of results will provide additional second-hand data to supplement their own counts.

It is not envisaged that the follow-up section should be used in a straightforward test situation, since guidance and discussion will no doubt be required along the way. Nevertheless, the eventual written responses may be used for assessment purposes with the use of a mark scheme.

Mark scheme

a	Explanation of sex inheritance		. . . 4
	Relate to data obtained		. . . 1
b	Assumption: wing length inheritance		. . . 1
	Genetic explanation of inheritance of wing length		. . . 2
c	Assumption: eye colour inheritance		. . . 2
	Genetic explanation of inheritance of eye colour		. . . 5
d	Genetic explanation of inheritance of wing length and eye colour together. The dihybrid cross.		. . . 8
	If assumptions correct, theoretical ratio obtained		. . . 2
e	Discussion of significance of probability level obtained from statistical analysis		. . . 4
f	Possible reasons for differences between practical and theoretical results	3 × 2	. . . 6
			35

Other ability which could be assessed

Calculations from data

The statistical analysis involving the chi-squared test may be used as a teaching situation and assessment can then be based on the quality of the final written work produced, possibly also taking into account contributions made by individuals to class discussion. Alternatively, if the theory and execution of such analysis have already been explained, then this section can stand as a simple test to be marked.

INVESTIGATION 8.4
The responses of woodlice to light and humidity

Equipment per student

- Stoppered tube containing 10 woodlice (of the same species and collected together). Alternatively, blowfly larvae may be used.
- 2 bottoms of small plastic Petri dishes (8–9 cm diameter), one with 1 cm diameter hole cut in centre
- Plasticine
- Surgical gauze
- Black card
- Cotton wool
- Adhesive labels
- Adhesive tape
- Anhydrous calcium chloride
- Stop clock
- Bench lamp
- Statistical tables for *t*–test

Fig 8.5 Woodlouse, dorsal view (×5)

Instructions given to students

1. Take the Petri dish without the hole and make a Plasticine division to divide it into two halves, A and B. The top of this division should be level with the top of the sides of the dish.
2. Stretch the gauze tightly across the top of this dish and hold it in position with small pieces of sticky paper on the outside.
3. Place the dish with the hole upside down over the first dish. Fasten the two together with sticky tape running round the junction. The upper dish forms Chamber C above the gauze, (Fig 8.6).

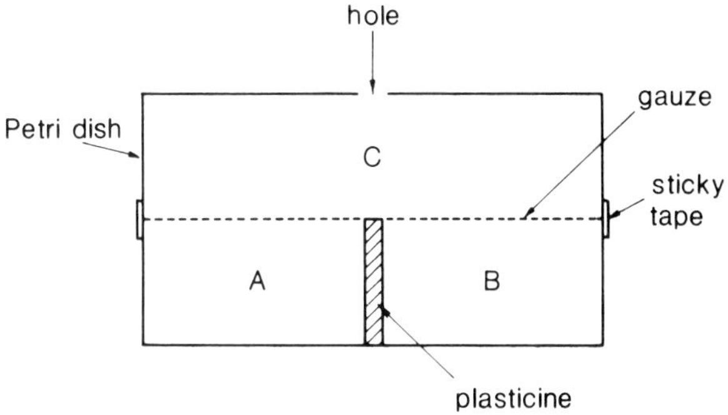

Fig 8.6 Section through the choice chamber apparatus

A Response to light

4 Drop 10 woodlice into Chamber C through the hole and cover the hole with a sticky label.

5 Place the piece of black card on top of the upper dish so that it covers Chamber A but not Chamber B.

6 Place the lamp directly over the apparatus at a height above it of about 30 cm.

7 After 1 minute, record the number of woodlice above Chamber A and above Chamber B.

8 Do not disturb the apparatus. Record again at 1 minute intervals until you have 5 sets of data. Tabulate results.

9 Dismantle the apparatus and place the woodlice back in the stoppered tube.

B Response to humidity

10 Place calcium chloride in Chamber A and damp cotton wool in Chamber B. Make sure that neither will touch the gauze above.

11 Replace gauze and upper dish as before and cover the hole with a sticky label. Make sure that the apparatus is fairly evenly illuminated (no light bulb nearer than about 30 cm). Leave for 3 to 5 minutes.

12 Remove label and drop 10 woodlice into Chamber C. Replace label.

13 After 1 minute record the number of woodlice above A and above B as before. Continue to record distribution at 1 minute intervals until you have 5 sets of data. Record results in a table.

C Response to light and humidity

14 Place the black card over the apparatus and illuminate with a lamp so that the woodlice have a choice between a dark, humid area and a light, dry area in Chamber C.

15 Leave the apparatus for 3 to 5 minutes. Record the distribution of woodlice at 1 minute intervals until you have 5 sets of data as before. Tabulate results.

Follow-up

1 Why is it important that the bench lamp is no nearer than about 30 cm from the apparatus?
2 What is the function of anhydrous calcium chloride?
3 Why, in parts *B* and *C* of the investigation, is it important to leave the apparatus for 3 to 5 minutes before adding the animals?
4 In part *C* of the investigation, over which Chamber did you place the black card?
5 What do the results from part *A* indicate about light preference of woodlice?
6 What do the results from part *B* indicate about humidity preference of woodlice?
7 From the results of parts *A, B* and *C* record the probable order of preference of the woodlice for the following:
dark and dry   light and dry   light and humid   dark and humid
8 *Statistical analysis*
Look again at the results for part *A* (or part *B*) of the investigation, the response to light. Statistical methods may be used to analyse results and help you decide if the environmental factor (light or humidity) is having a significant effect on activity of the woodlice and therefore on their distribution.

a A null hypothesis is set up: assume that light has no effect on distribution and so in this case equal numbers of animals should be found on either side of the dish.

b Calculate the standard error (SE) which reflects the distribution which would be obtained if chance alone operated.

$$\text{SE} = \sqrt{\frac{\text{expected proportion}}{\text{number of animals in trial}}}$$

Let p,q be the expected proportions where $p + q = 1$; $n =$ number of animals.

Since the 50 animals were equally distributed,
$p = 0.5,\ q = 0.5$ and $n = 50$

$$\text{SE} = \sqrt{\frac{pq}{n}} = \sqrt{\frac{0.5 \times 0.5}{50}}$$

c From the counts made calculate the observed ratio.

d Calculate the difference between the observed and expected ratios.

e These results can now be checked using statistical *t* tables.

$$t = \frac{\text{difference between observed and expected ratio}}{\text{standard error}}$$

Substitute into this and obtain a value for *t*.

f Two further factors now need to be decided.
Degrees of freedom: this equals the total number of observations minus one. In this case, 49.
Acceptable probability level: for most biological work a probability of 5% (0.05) is considered to be significant.

Table 8.3 Distribution of t

Degrees of freedom	Probability, *p*				
	0·1	0·05	0·02	0·01	0·001
1	6·31	12·71	31·82	63·66	636·62
2	2·92	4·30	6·97	9·93	31·60
3	2·35	3·18	4·54	5·84	12·92
4	2·13	2·78	3·75	4·60	8·61
5	2·02	2·57	3·37	4·03	6·87
6	1·94	2·45	3·14	3·71	5·96
7	1·89	2·37	3·00	3·50	5·41
8	1·86	2·31	2·90	3·36	5·04
9	1·83	2·26	2·82	3·25	4·78
10	1·81	2·23	2·76	3·17	4·59
11	1·80	2·20	2·72	3·11	4·44
12	1·78	2·18	2·68	3·06	4·32
13	1·77	2·16	2·65	3·01	4·22
14	1·76	2·14	2·62	2·98	4·14
15	1·75	2·13	2·60	2·95	4·07
16	1·75	2·12	2·58	2·92	4·02
17	1·74	2·11	2·57	2·90	3·97
18	1·73	2·10	2·55	2·88	3·92
19	1·73	2·09	2·54	2·86	3·88
20	1·72	2·09	2·53	2·85	3·85
21	1·72	2·08	2·52	2·83	3·82
22	1·72	2·07	2·51	2·82	3·79
23	1·71	2·07	2·50	2·81	3·77
24	1·71	2·06	2·49	2·80	3·75
25	1·71	2·06	2·49	2·79	3·73
26	1·71	2·06	2·48	2·78	3·71
27	1·70	2·05	2·47	2·77	3·69
28	1·70	2·05	2·47	2·76	3·67
29	1·70	2·05	2·46	2·76	3·66
30	1·70	2·04	2·46	2·75	3·65
40	1·68	2·02	2·42	2·70	3·55
60	1·67	2·00	2·39	2·66	3·46
120	1·66	1·98	2·36	2·62	3·37
∞	1·65	1·96	2·33	2·58	3·29

g Look in the *t* tables (Table 8.3) and find the critical value of *t* with 49 degrees of freedom at the 5% significance level. If the calculated value of *t* is greater than the critical value, at this 5% significance level the null hypothesis may be rejected.

h Record the critical and calculated values of *t*. Which is the greater? Can the null hypothesis be rejected?

i Is it therefore safe to assume that at·this 5% significance level there is a real difference in distribution of woodlice which cannot be accounted for by chance alone?

Assessment

Checklist

1 Choice chamber apparatus made according to instructions.
2 Procedure for investigation followed closely and thoughtfully.
3 Results recorded in tables for parts *A* and *B*.
4 Apparatus assembled correctly for part *C*. Results tabulated.
5 Data interpreted in response to structured questions.
6 Statistical calculations performed correctly. Deductions made in response to data produced.

This is a lengthy investigation and will probably not be completed in one session. The range of woodlice response is always great and interpretation may be a more worthwhile exercise if it is based on the class results as a whole.

The data is suitable for statistical analysis and provides a good introduction to the use of such techniques and the value they have in drawing conclusions from the results tabulated.

Mark scheme

a Deductions made from results of part *A* . . . 4

b Deductions made from results of part *B* . . . 4

c Order of preference based on observations made in parts *A*, *B* and *C* . . . 8

d Statistical analysis:
Rejection of null hypothesis . . . 2

Statement: at 5% significance level, difference in distribution due to light . . . 2

20

Other abilities which could be assessed

Manipulative skills

Although ready-made choice chambers may be used this exercise involves the assembling of the apparatus as well as the handling of the animals. In view of the fact that it may take more than one session to complete, some of the time could be given to an assessment of manual dexterity.

Following instructions

Questions 1–4 of the follow-up section are designed to test the students' comprehension of the practical procedure. The marking of this written work, along with observations and questioning of the students at work, will provide the necessary information for an assessment of this ability.

Presentation of results and calculations from data

If the exercise and the data collected are to be used as an introduction to statistical analysis then the t test may well be carried out as a teacher-directed class task. Alternatively, with further information provided and individual help given, students may tackle the calculations alone. Either way they can be ranked on the quality of work produced.

Chapter 9

EXPERIMENTAL DESIGN

Whether the experiment which is designed by the student is also to be carried out, or whether it is a theoretical exercise only, the stages involved in the assessment remain the same. The student must show an ability to recognise a problem, formulate a hypothesis, devise a logical and timed work plan and, choosing appropriate equipment and techniques with suitable controls, test the hypothesis. Finally, he or she should be able to choose a suitable method of presenting the results obtained and, by reference to the initial hypothesis, draw meaningful conclusions from them.

Experimental design

Many teachers involved in assessment would agree that the ability to plan and carry out investigations stands apart from other abilities and probably requires special treatment. It involves a more demanding thinking skill than tested elsewhere, and for students and teachers alike it poses problems.

Assessment of the ability involves placing students in situations which are novel and problematical while higher intellectual demands are made of them. Some are anxious, work with less confidence and ultimately perform less well. Teachers have difficulties both in finding material which they can rely on for valid and objective testing and also in organising the assessment procedure itself. These problems are very real and research has shown that, compared with the other skills, fewer assessments are carried out, they are left until later in the course and lower marks are achieved. However, this ability discriminates well between students and its inclusion is justified on the grounds that a worthwhile science course should be providing opportunities to practise the development of such an essential scientific skill.

Although experimental design may be part of the follow-up work to other investigations, for assessment purposes it is more convenient to use separate exercises. Students may not always be asked to carry out the practical procedure, but the planning stage will involve the same well-defined structured sequence, and a mark scheme for the assessment of experimental design can be based on this checklist.

Checklist with marks

1 Recognition of the problem and a statement of the hypothesis under investigation . . . 2

2 A clear specification of the purpose of the investigation . . . 1

3 Choice of an appropriate practical procedure with the production of a work plan. This will include:

a List of apparatus, materials and chemicals . . . 3

b Correct sequence and timing of the series of operations involved such that someone else could carry it out . . . 5

c Recognition of the shortcomings of the practical with sources of error, limitations of data and need for controls considered . . . 3

d Details of the ways in which the results obtained may be processed and analysed . . . 4

e Appreciation of the extent to which the results will substantiate the original hypothesis . . . 2

4 Efficiency at completing the practical procedure . . . 5

25

There are a number of different approaches to the assessment of experimental design. The following sections outline seven methods giving several suitable assessments for each.

METHOD 9.1
Extension or modification of a completed investigation

A simple yet very effective way to introduce the idea of planning an investigation is to ask the student to extend, modify or refine a practical procedure already completed. The student, having recent working experience of the laboratory techniques involved, will be in a good position to suggest a viable work plan for solving the new problem. Although this is experimental design on a small scale only, it is found by teachers to be a useful and valid method. It is well within the grasp of most students and is a valuable training exercise in a difficult skill.

Examples
The follow-up work and assessment sections of Investigations 4.2, 5.2, 5.3, 6.3, 6.4, 7.2, 7.4 and 8.1.

METHOD 9.2
Written plans

Another straightforward approach, as far as the assessor is concerned, is to provide the student with information and a problem and ask for a written plan only. There is no shortage of material, no resources are required and only a relatively short time allocation. It will be evident, however, that students cannot be expected to tackle purely theoretical problems without being able to draw on a fund of past experiences and practice which will guide them in how to proceed. For this reason such an approach will be possible only after a course has been in operation for some time, and even then weaker students may continue to have difficulties.

Examples

1 Vitamin B_{12} is a growth factor required by the unicellular organism *Euglena.* In pernicious anaemia in man the body is deficient in vitamin B_{12}. Given separate samples of serum from two different patients suffering from pernicious anaemia, how could you use *Euglena* to compare the extent to which each was deficient in vitamin B_{12}? (JMB)

2 Deterioration of stored grain is an important economic problem. How would you test the hypothesis that there is a correlation between the moisture content of wheat grains and their respiratory rate? (JMB)

3 Investigations into the germination of pollen grains of a plant involved growing the pollen grains in small drops of a nutrient medium. The drops were of uniform volume, but the number of grains in each drop differed. The results obtained are given in column A of Table 9.1.

Table 9.1 Results of investigations into the germination of pollen grains

Approximate number of pollen grains per 0.01 cm^3 drop	Percentage germination of pollen grains	
	A	B
10	3	73
20	9	74
40	13	73
60	19	79
80	31	73
100	42	81
150	60	*
200	63	*
250	72	*
300	78	*

* Results were not recorded

When the investigation was repeated using the same medium except for the addition of pollen grain extract to it, the results in column B of the table were obtained.

a Plot the results from both investigations on one graph.

b It might be thought that competition for nutrient materials influenced the germination of the pollen grains. How can this suggestion be refuted from the results?

c Is it likely from the results that an environmental factor such as oxygen is limiting the germination of the pollen grains?

d Suggest a hypothesis to account for the following results:

- *i* When the number of pollen grains in a drop is high, a relatively high proportion germinates.
- *ii* When extract is added to the drops, the proportion of grains which germinates is high no matter what the pollen grain population density is.

e Using the following headings, explain briefly how you would test your hypothesis:

- *i* Methods employed.
- *ii* Results expected if your hypothesis was correct.
- *iii* How these results would support the hypothesis.

Assessment of Example 3

The plotting of the graph and its interpretation can be marked and used to test the appropriate skills. Parts *d* and *e* of the question are concerned with experimental design. The mark scheme covers these sections only and suggests one possible solution to the planning of experimental detail, although there will be a variety of responses from students.

Mark scheme

1 *Hypothesis*

Pollen grains secrete a chemical which diffuses through the nutrient medium . . . 1

Chemical stimulates germination of other pollen grains . . . 1

More pollen grains produce a greater concentration of chemical per unit volume of medium. Therefore germination rate is greater . . . 1

Pollen grain extract contains the chemical. Therefore germination rate is high regardless of density of grains . . . 1

2 *Purpose of the investigation*

Increasing concentrations of pollen grain extract are added to a controlled experiment, and the effect on the rate of pollen grain germination is observed . . . 1

3 *Work plan*

a *Equipment required*

Standard nutrient medium
Preparation of pollen grain extract at increasing concentrations: 0%, 0.1%, 1%, 10%
Water bath with temperature control
Cavity slides
Cover slips
Pipettes
Microscope with light source . . . 3

b *Sequence of operations*

i Constant volume of nutrient medium
Known steady incubation temperature
Constant number of pollen grains
Known volume of distilled water (0% extract) added
Incubate for a known time
Count and record number of pollen grains germinating

ii Repeat procedure exactly, but add equal volume of 0.1% pollen grain extract instead of distilled water
iii Repeat with range of concentration of extract
Count and record germination of the grains . . . 5

c *Limitations of procedure*
Maintenance of constant factors
Identification of grain germination
Accuracy of small-scale work . . . 2

d *Results*
Calculation of germination rate
Plotting of a graph: rate of germination against concentration of pollen grain extract added
Sketch of shape of graph obtained, indicating a higher germination rate for increasing concentrations of extract added . . . 3

e Statement of how the results substantiate the hypothesis . . . 2

20

METHOD 9.3
Individual planning and completion of an investigation

It is possible to present a problem to students and then, having supplied a range of relevant apparatus and chemicals, give them complete freedom to plan and execute the investigation in the laboratory. Usually, the students will have had prior experience of a technique and will be expected to apply it to a new situation. Some will respond well to, and benefit from, the challenge of this more open-ended approach. Poorer students may have difficulty in coping with a situation in which no instructions are given and in which they initially have to draw on their own resources. Teachers will need to prompt, guide and direct them to a greater degree. Such students will of course receive a lower mark than those who carry out a satisfactory practical procedure based on an acceptable approach which they have formulated for themselves.

The number and types of investigation possible using this approach are limited and the method can be expensive and time-consuming.

Examples

1 Devise a method to estimate the field of view of the microscope. Then compare the stomatal frequency on the upper and lower epidermis of the leaf provided.

2 You are provided with a standard DCPIP solution and an ascorbic acid (vitamin C) solution of known concentration. Using these and any other simple apparatus, estimate the vitamin C content of the fresh and preserved fruit extracts provided.

3 You are given a mixture of a crystalloidal (sodium chloride) solution and a colloidal (starch) solution. Devise a means to separate and identify these two components.

Assessment of Example 3

An outline mark scheme is given for the assessment of the initial planning stage and of the student's ability to carry out the investigation. A more detailed version has been used to grade the two examples of students' written work given below.

Both students have selected the same suitable technique with correct tests, procedure and sequence recorded. However, not only does student B include more detail of experimental method such as length of tubing used and time allowed, he also appreciates the need for refinements such as washing the Visking tube initially and using methods to maintain a concentration gradient. This candidate shows full comprehension of the problem and an excellent ability to devise a sound scheme of work.

Mark scheme

1	Explanation of the principle of the method of separation	. . . 3
2	Statement of tests used to identify the components	. . . 2
3	Details of experimental method for separation	. . . 10
4	Final results of tests for the two components after separation	. . . 5
		20
5	Efficiency in completing the practical procedure	. . . 5

Details of the mark allocation for students A and B are given in Table 9.2.

Table 9.2 Allocation of marks

Mark scheme section	Marks for Student A	Marks for Student B
1	2	3
2	1	2
3	4	10
4	2	5
Total	9	20

Student A
One end of a piece of Visking tubing was sealed with a piece of cotton to make a small bag. Into this was put a sample of the given solution. The other end of the bag was fastened off. The Visking tubing was placed in a beaker of water and left for a short time to allow dialysis to occur. The bag was then taken out of the water and placed to one side.

From the beaker two samples of water were removed. One was tested for chloride ions with silver nitrate solution and the other was tested for starch using iodine solution. This was to show that only the chloride ions were present in the water, as starch molecules were too large to pass through the Visking tubing.

A sample of mixture from inside the bag was taken and tested with silver nitrate solution to show chloride ions were not present. A separate sample was tested with iodine to show that starch was present.

Mark: 9/20

Student B
A 20 cm length of narrow Visking tubing was cut and one end was tied off tightly with a piece of cotton. The prepared mixture was carefully pipetted in until the tube was almost full and then the open end was sealed with cotton.

The tubing was washed to remove solution from the outside and it was then completely submerged in a beaker of distilled water. It is important to use distilled water. The small sodium chloride molecules can diffuse out from the high concentration in the bag to the low concentration in the distilled water. The starch molecules are too large to pass through the semi-permeable membrane and so stayed inside the bag. From time to time the bag was stirred round to distribute sodium chloride molecules more evenly and so assist the concentration gradient.

After 20 minutes the bag was transferred to another beaker of distilled water. In the first beaker the concentration gradient for sodium chloride molecules would be decreasing. The water in the first beaker was tested for chloride ions using silver nitrate solution which in this case gave a white precipitate of silver chloride. The sample tested for starch with iodine in potassium iodide solution did not give a blue-black colour.

The bag was transferred to a fresh sample of distilled water at 20 minute intervals. In each case the water sample was found to be positive for sodium chloride but negative for starch.

Finally, when the sodium chloride test was negative, it was assumed that all the sodium chloride had diffused out of the tubing. Tests carried out on the contents of the Visking tubing revealed that starch was present but not sodium chloride.

Mark: 20/20

METHOD 9.4
Individual planning with prepared schemes for the practical work

A well tried and trusted way to assess experimental design is to present the problem to students who then, individually and without assistance, design a work plan for the investigation. The work plan is assessed, although the student is then provided with a teacher-prepared plan for the actual practical procedure.

A very successful refinement of this technique involves the teacher in discussing, amending and elaborating the original plan with the student so that both come to an agreement on the most suitable method, as well as on the assessment of the plan. The student then completes the exercise.

The method works well, since the individual's original ideas are combined with help and direction from the teaching staff. The approach is an effective way of training in the skill of experimental design as well as being useful in assessment.

Examples

1 Devise a method to investigate the effect of temperature on the production of carbon dioxide by yeast.

2 When beetroot is cooked, the surrounding water becomes red. How would you test the hypothesis that the release of red pigment from beetroot does not have a linear relationship with temperature? (JMB)

3 Plan and carry out an investigation into the distribution of protein and starch digesting enzymes in the gut of an earthworm.

Assessment of Example 3

The individual experimental design plans can be assessed by reference to a mark scheme. A worksheet is then provided and may either be used directly or may act as a guide in modifying the student's scheme. Either method will enable the student to carry out the investigation so that assessment of practical skill, analysis of results and discussion of the limitations of the technique can take place.

Mark scheme

1	Details of the dissection technique	. . . 2
2	Identification and removal of sections of the gut	. . . 3
3	Maceration of the gut to release enzymes	. . . 1
4	Method used to test for proteases Timing. Recording of results	. . . 3
5	Method used to test for amylases Timing. Recording of results	. . . 3
6	Presentation and discussion of results	. . . 4
7	Discussion of limitations of techniques and suggestions for improvements of procedure	. . . 4
8	Efficiency in carrying out the practical investigation	. . . 5
		25

Worksheet

Equipment per student

Freshly-killed earthworm (immerse in cooled, boiled water)
5 cm^3 0.5% soluble starch solution
Distilled water
Iodine in potassium iodide solution
Dissection dish
Dissection instruments
5 cm^3 graduated syringe
8 watch glasses
8 glass rods
4 spotting tiles
4 micro pipettes
Clinistix strips
4 wooden splints
Strip of photographic negative

Instructions given to students

1 With the dorsal surface of the worm uppermost, make an incision, well behind the clitellum, into the body wall. Cut along the mid-dorsal line to the anterior end, keeping the points of the scissor blades up.
2 Insert a pin through the anterior end and anchor in the dissection dish. Pull gently backwards and pin through the posterior region. Pin out the body wall flaps to left and right, snipping through septa as necessary.
3 Identify each of the four sections of the gut as indicated on the diagram Fig 9.1 and carefully remove into four separate labelled watch glasses.
4 To each add 2 cm^3 distilled water and thoroughly chop up the gut wall and contents. Mix well and divide this solution between two watch glasses, A and B.

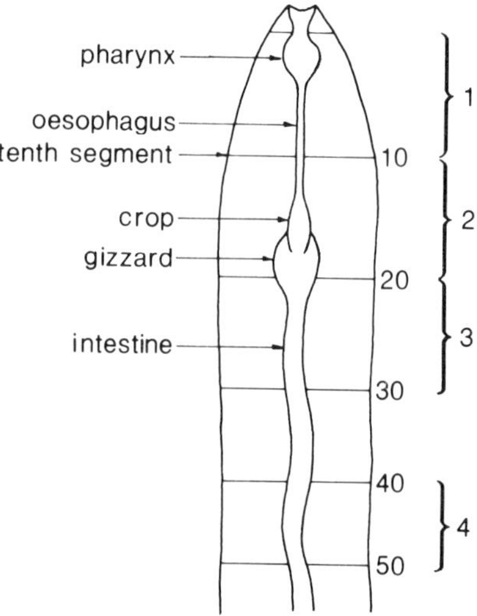

Fig 9.1 Sections 1–4 of the earthworm gut for enzyme analysis

5 For each section of the gut place a strip of photographic negative attached to a splint into watch glass A and add an equal volume of the starch solution to B. Note the time.

6 At intervals of 5 minutes and for up to 1 hour:

   a Examine the negative strip in A. Record results and note the time taken for the emulsion to digest and the film to become clear.

   b Remove with a pipette a drop of solution from B. Test with iodine solution on a spotting tile. Record results and note the time taken for a negative starch test.

   c Test the solution in B with a Clinistix strip. Record results and note any colour change from pink to blue.

7 Tabulate all results.

8 Discuss the distribution of the enzymes.

9 Note any limitations of the procedure and suggest modifications to improve the experimental technique.

METHOD 9.5
Class discussion followed by individual work

Even at the highest levels, research is rarely carried out in isolation, and experimental design lends itself well to class discussion and teacher-student consultation methods. A problem is presented to the class and by discussion and teacher direction a common and acceptable work plan is produced. Individuals may then be left to carry out the procedure. Students who find difficulty in constucting an effective experimental design when working alone are often stimulated to put forward original and useful contributions in a group where they can draw on the experience of others as well.

Although the assessor has the difficult task of identifying and grading individual contributions, group work is valuable and stimulating and well worth operating at some stage. This approach also has the advantage that more complex investigations than normal may be considered.

Examples

1 How would you test the hypothesis that concentrations of DDT which are non-lethal to the aquatic plant *Elodea* do limit its photosynthesis? (JMB)

2 Some smokers have saliva which is less effective in digesting starch than is the saliva of non-smokers. How would you test the hypothesis that cigarette smoke inactivates the enzyme amylase? (JMB)

3 How would you test the following hypothesis: the rigidity of potato tuber tissue is proportional to its water content?

Assessment of Example 3

It is possible to rank and grade students during or after a class discussion session, and a list of criteria for the award of marks is given. Reliability of assessment will depend on factors such as the size of the group and the personality of individuals within it. The presence of a colleague is helpful.

An example of a worksheet for the investigation is also provided. Although it may be modified during the group work, the teacher will be more effective in structuring the discussion if such a prepared plan is available. Finally, a mark can be allocated for the efficiency with which the student carries out the procedure.

Mark scheme

Shows an immediate and lively response to the question posed. Appreciates fully the purpose of the investigation and has well-structured and original ideas on techniques to be used. Able to think in specific and detailed terms about the sequence and timing of the relevant method.	. . . 9–10
Shows clear evidence that the problem is understood and contributes thoughtfully to the discussion of practical detail	. . . 7–8
Slower to grasp the aims of the experimental method and lacks imagination over planning the steps in the procedure. However, with help begins to develop some ideas of a basic plan.	. . . 5–6
Does not really appreciate either the nature of the investigation or the basic procedure to be adopted. Has difficulty thinking in practical terms about the experiment.	. . . 3–4
Fails to understand the problem completely. Unable to contribute any useful ideas on the practical techniques to be employed.	. . . 1–2

Worksheet

Equipment per student

Potato

50 cm^3 100% salt solution

Oven at 100°C

Top-pan balance

Scalpel

Cork borer

Measuring cylinder

6 Petri dishes with lids

Filter paper

Specimen tube with cork

Dressmaking pins

Graph paper

Protractor

Instructions given to students

1 You are provided with a saturated salt solution. Using this and dilutions you will make from it, half fill each Petri dish with one of the following:
   a 100% salt solution (saturated)
   b 30% salt solution
   c 15% salt solution
   d 7.5% salt solution
   e distilled water
   Label each dish.
2 Peel the potato and prepare five chips of identical size (cork borer size 2 or 3). Place all chips in distilled water for 10 minutes. Remove and blot gently.
3 Put one chip into each of the Petri dishes *a* – *e*. Note the time and leave for 30 minutes.
4 Measure the degree of rigidity of each chip using the apparatus shown in Fig 9.2. Record the final position of the marker pin directly on to graph paper fixed behind the apparatus. Draw a line between the position where the chips were anchored and the mark on the graph paper. Measure the angle between this line and the horizontal.

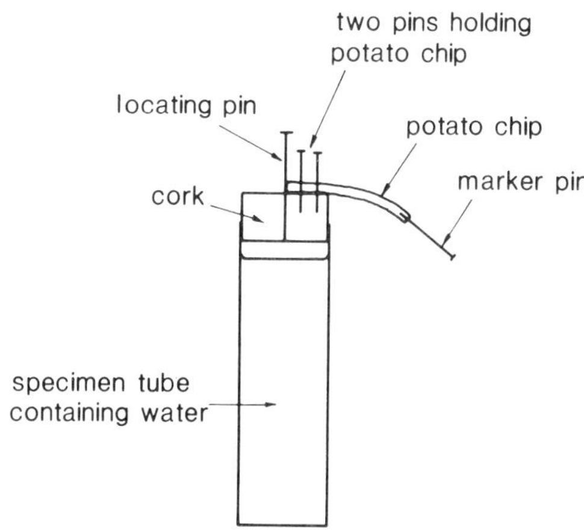

Fig 9.2 Apparatus for recording the rigidity of a potato chip

5 Estimate the water content of each chip by weighing initially and then heating in an oven at 100°C to constant mass. Calculate the percentage water content of each chip.
6 Tabulate results.
7 Plot a graph of the degree of bending of each chip against water content.
8 Discuss the shape of the graph.

METHOD 9.6
Field work

Field work often provides an excellent opportunity to assess planning skills. Initially students familiarise themselves with the flora and fauna of the area and learn the techniques of ecological studies. Problems which can be investigated in the field may become apparent at this stage. If not, teachers may have to provide a list of suitable projects. Although the environment is different and the time allocation longer, the design stage follows the same steps given in the checklist with marks at the beginning of the chapter. This can then be used for assessment. Students will generally work in groups, which gives all the advantages of shared resources along with inherent problems for the assessor.

Examples

1 The larva of the mayfly lives in fresh water. It has along its abdomen a series of plate-like gills which vibrate up and down and provide surfaces for gas exchange. How would you test the hypothesis that the rate of gill movement is inversely proportional to the oxygen content of the surrounding water? (JMB)

2 Design an investigation to consider the effect of the oxygen content of a stream on the distribution of crustaceans there.

3 Design a project to investigate the distribution of clover colonies on the school field.

METHOD 9.7
Laboratory Projects

Assessment of experimental design is not generally to be interpreted as the extent to which a candidate can undertake an independent research project. Nevertheless, many teachers allow investigative projects to be undertaken and if resources and time allow, such a project provides good material for the assessment of experimental design. However, while there is much to be gained in many ways from long term and detailed individual work, a range of shorter investigations allows more opportunities for testing.

Examples

1 How could you test the hypothesis that, in an insect with a complete life history, the rate of metamorphosis is influenced by temperature? (JMB)

2 Design an investigation to locate the region of a broad bean root which is responsible for gravity reception. (JMB)

3 Goldfish can be trained to swim through a maze. How would you test the hypothesis that the rate of learning in goldfish is directly proportional to the temperature of the surrounding water? (JMB)

Chapter 10

FURTHER ASSESSMENT EXERCISES

The following investigations are taken, with permission, from past Advanced level biology practical papers of GCE boards.

They have been selected as being useful for internal assessment. For each an equipment list is provided and an indication given of the skills which may be tested. Further reference to this source of material will provide many similar examples.

Investigation 1

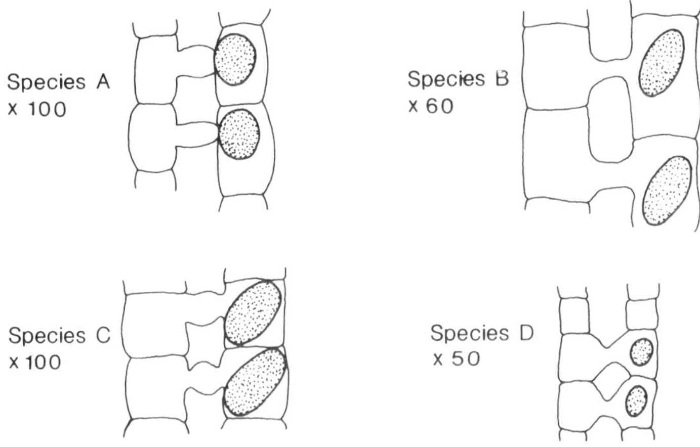

Fig 10.1 Four species of *Spirogyra*

The diagrams in Fig 10.1 show four species of *Spirogyra* as viewed under a compound microscope. The magnification used for each species is stated.

Measure the width of the filament at the transverse wall in each species.

Using the magnification given beside each diagram, calculate the actual widths of the filaments.

Present your answers in the form of a table. (AEB)

Equipment per student
Ruler (millimetre)

Ability which could be assessed
Presentation of experimental results with calculations

Investigation 2

Specimens S4, S5 and S6 are from plants bearing spiny projections which differ in their origin and position.

a Examine the specimens carefully and, by means of a labelled drawing of each, show the form and arrangement of the spines.

b Cut a length (approximately 3 cm) of stem from S5, selecting a part which includes a rigid spine. Hold the piece of stem vertically on the bench. With a sharp scalpel cut the stem in half longitudinally passing through a spine and its base. Repeat this procedure with specimen S6.

c Place the cut lengths of stem of S5 and S6 cut face downwards into phloroglucinol in a watch glass. After a minute, transfer the pieces to a clean watch glass and add a few drops of concentrated hydrochloric acid (CAUTION) to the cut surfaces. After a few minutes, lignified tissue will stain red.

d Using forceps, transfer the pieces of stem on to filter paper and blot them (CAUTION). Examine the cut, stained faces with a hand lens. Draw diagrams to show the distribution of lignified tissues in the stem and spine of S5 and S6.

e Answer the following, using your observations of intact and sectioned specimens to support your conclusions.
  i What structure has been modified to form the spines in S4 and S5?
  ii How does the distribution of spines in S5 differ from that in S6?
  iii How does the origin of the spines in S5 differ from that in S6? (JMB)

Equipment per student
S4 Length of stem with at least two leaves of *Berberis* (barberry) with spines and leaves.
S5 Length of stem with at least two leaves of *Crataegus* (hawthorn) with some large rigid spines and leaves.
S6 Length of stem with at least two leaves of *Rosa* (rose) with reasonably sized firm spines and leaves.

Phloroglucinol solution
Concentrated hydrochloric acid
Distilled water
Microscope slides
Cover slips
Filter paper
Watch glasses
Dissection instruments

Abilities which could be assessed
Manipulative skills
Observation, identification, recording and interpretation

Investigation 3

a Determine, in as much detail as possible, the differences in *external* features between the gynoecium (carpels) of a flower from K2 and the gynoecium of a flower from K3.

From the inflorescence K2 select a single flower in which the stigma protrudes as much as possible. Using a sharp needle, remove this flower in its entirety and place it on a slide. With a pair of needles, dissect away and discard all the structures surrounding the gynoecium, leaving only the ovary with the stigma and style on the slide.

In a similar manner, on another slide, obtain the complete gynoecium from a flower of the plant K3.

i Examine these two structures carefully with a lens and make a very large accurate drawing of each, to the *same* scale, to show the relative sizes and shapes of the various regions. Do not draw any details of the regions.

ii State the magnification of your drawings.

iii On each drawing show clearly the overall dimensions (to the nearest 0.5 mm).

iv Now cut through the junction of the style and ovary of each of the two gynoecia and discard the ovaries, leaving only the style and stigma(s) on each slide. Add two drops of the stain methylene blue to each and cover with a cover slip.

Make a large drawing of a very small portion of a stigma of each plant to show accurately the external features and any structures adhering to them.

b Carefully observe specimens K4, K5, K6 and K7 and answer the following questions with regard to their external features only.

i State two taxonomic features which are common to K5, K6 and K7 but not seen in K4.

ii State three further features which are common to K6 and K7 but not seen in K5.

iii State three differences between K6 and K7.

iv Using your observations and the sequence in *i* to *iii* construct a simple dichotomous key which could be used to distinguish the four specimens.

(C)

Equipment per student

K2 One inflorescence of plantain in water (young inflorescence with mature stigmas protruding from the lower flowers)
K3 Two or three flowers of stitchwort in water
5 cm^3 0.1% aqueous methylene blue
Microscope slides
Cover slips
Teat pipette
Dissection instruments
Hand lens
K4 Preserved nematode
K5 Freshly killed earthworm
K6 Preserved caterpillar
K7 Preserved woodlouse

Abilities which could be assessed

Manipulative skills
Observation, identification, recording and interpretation

Investigation 4

a *i* Make a fully labelled drawing of specimen P.
ii Indicate clearly on your drawing the positions and types of joint present in P.

b *i* Identify specimen Q.
ii List *five* differences between specimens P and Q.

(L)

Equipment per student

P Hind limb skeleton of small mammal (not including pelvic girdle)
Q Third leg of adult locust

Abilities which could be assessed

Observation, identification, recording and interpretation

Investigation 5

The preparations N13 and N14 are transverse sections of two tubular structures found in different parts of a mammal's body. Examine each section with a hand lens and under the low and high powers of the microscope. Make simple line diagrams showing the relative proportions and main components of the different layers forming the walls of these structures. Do *not* draw cells. Label your drawings to indicate

a the nature of the lining, and
b the main tissue components of the wall.

From what organs do you think these sections were made? Give reasons for your answers based on your own observations.

(JMB)

Equipment per student

N13 Prepared microslide TS rat trachea with ciliated epithelium and cartilage
N14 Prepared microslide TS rat ileum with villi
Microscope with light source
Hand lens

Abilities which could be assessed

Observation, identification, recording and interpretation

Investigation 6

a Dissect the mammal to show the vessels through which oxygenated blood passes from the lungs to one side of the head. The vessels should be cleanly dissected and freed of surrounding tissues. Make a labelled drawing of your dissection.

b Flag label the following:

A An organ in which deamination occurs.
B A region of the gut from which a histological section would reveal columnar cells, goblet cells and Brunner's glands.
C *Either* the tube in which a placenta may be formed
or the duct through which the contents of vasa efferentia are passed to the vas deferens.
D An organ which can produce a substance preparing the body for action in emergencies and increases cardiac frequency.

(OC)

Equipment per student

Freshly killed or preserved small mammal
Dissection board and pins
Dissection instruments
Cotton
Pins
Labels

Abilities which could be assessed

Manipulative skills
Observation, identification, recording and interpretation

Investigation 7

You are provided with a male and a female cabbage white butterfly. The female is distinguished from the male by the possession of two additional black spots on the upper surface of each forewing. On each wing of both butterflies there is quite a large area (labelled A on Fig 10.2) across which the veins do not run.

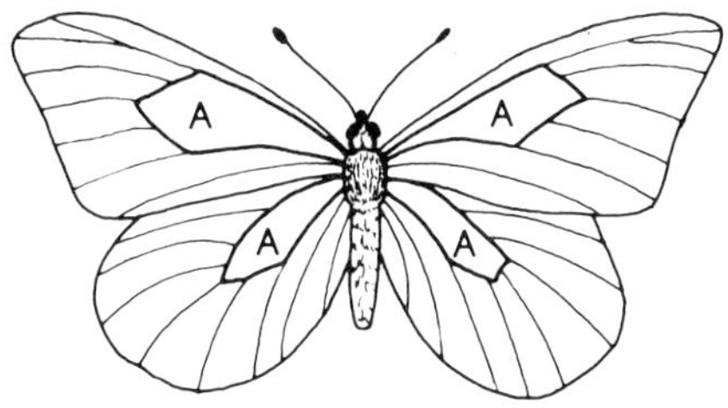

Fig 10.2 Reference diagram: wings of the cabbage white butterfly

a Take the female butterfly and, with the point of a scalpel, gently scrape some of the scales from area A on the upper surface of one of the forewings and transfer them to a microscope slide.

b Wipe the scalpel thoroughly and then, in a similar manner, place some of the scales from area A on the upper surface of one of the hindwings on to another slide.

c Examine the scales under a microscope and draw a typical scale from each of these areas. Note that some scales have several teeth at their free edges, whilst others end in a single point or blunt end (one tooth).

d Now place some scales from the black wing tip of the female onto another slide.

e From each of the three regions, select samples of 20 scales at random. Count the number of teeth on the free edge of each of these scales and tabulate your results. For each region find the most frequently occurring number of teeth per scale and also the mean number of teeth per scale.

f Take a sample of scales from area A of the upper surface of the forewing of the male. Draw an example of the most frequently occurring type of scale from this area and list the ways in which it differs from the most frequently occurring type of scale in the corresponding area of the female.

(JMB)

Equipment per student

Male cabbage white butterfly
Female cabbage white butterfly
Scalpel
Microscope slides
Microscope with light source

Abilities which could be assessed

Manipulative skills
Observation, identification, recording and interpretation
Presentation of results with calculations

Investigation 8

Investigate some of the anatomical features of the insect provided.

a Cut off the head and pin it to a board with its ventral (posterior) surface uppermost.

Hold the body of the insect in one hand and, using a pair of sharp scissors, cut through *only* the exoskeleton in the mid-dorsal line, along the whole length of the abdomen. Place the insect in the dissection dish and, with pins, gently pull the cut edges apart. Pin the insect to the dish so as to expose the contents of the abdominal cavity. Cover the insect with saline solution. With a blunt seeker, gently move the structures to reveal small, white, glistening structures (parts of the tracheal system). With forceps and scissors remove a piece of trachea, about 1 cm long, and place it on a clean slide in a few drops of saline. Cover it with a dish to prevent it from drying.

Gently free the alimentary canal along its whole length to expose the hepatic (digestive) caeca, immediately behind the gizzard, and the Malpighian tubules originating between the mid and hind gut. With scissors and forceps remove two or three hepatic caeca and place them on a slide in a few drops of saline. Similarly, place a few Malpighian tubules on a clean slide in saline.

Observe these three structures carefully, using a hand lens and your microscope, and draw up a table of all the differences you can observe between these structures.

Using filter paper, remove the saline from each slide and cover each structure with about 5 drops of the stain borax carmine. Cover the slides to prevent them drying. Record the time and leave them for 10 minutes.

b During this 10 minutes, using forceps and the tip of a very sharp scalpel, remove in their entirety

i the labium (lower lip), and

ii the right maxilla from the head.

Place these two structures on the diagram Fig 10.3 in their correct positions. Fix the structures to the paper by covering them with transparent adhesive tape.

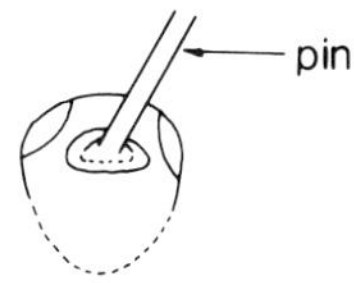

Fig 10.3 Outline of insect's head, ventral (posterior) view

c Using filter paper remove all the excess stain from the three slides. Cover each structure with acid alcohol. Leave for 2 minutes and then, with filter paper, remove the acid alcohol and replace it with 70% alcohol. Leave for two minutes and then replace the alcohol with dilute glycerol solution. Cover the Malpighian tubules and the piece of trachea with cover slips and observe them under the high power of your microscope. Make a large, labelled high power drawing of a very small but typical portion of a Malpighian tubule and of the piece of trachea.

d Draw up a table of further observable differences between these two structures resulting from the staining technique. State briefly why different parts of these structures stain differently with the same stain.

(C)

Equipment per student

Locust or cockroach
Insect saline
Borax carmine
70% alcohol
Acid alcohol
Dilute glycerol solution
Waxed dish
Microscope slides
Cover slips
Pins
Teat pipette
Filter paper
3 Petri dish lids
Wax pencil
Adhesive tape
Dissection instruments
Hand lens
Clock
Microscope with light source

Abilities which could be assessed

Manipulative skills
Observation, identification, recording and interpretation

Investigation 9

Triphenyl tetrazolium chloride (TTC) forms a colourless solution in water. The compound can act as a hydrogen acceptor and when reduced it forms an insoluble red compound.

a Take the broad bean seedling supplied and remove the testa carefully. Slice the seedling longitudinally into two equal portions, cutting first through the plumule and then the radicle before separating the two cotyledons.

b Boil one half in water for 3 minutes, then cool.

c Place each half in a separate labelled container and add sufficient 0.5% TTC solution to cover the specimen. Set aside for 30 minutes.

d Using forceps remove the specimens from the solution, place them in separate labelled containers and cover with water.

e Examine both halves of the treated seedling and make fully labelled drawings to show the structure of the seedling halves. Indicate on your drawings the location and intensity of the red stain. Record the length of the radicle.

f Explain the results you obtain. (W)

Equipment per student

Broad bean seedling with plumule and radicle
0.5% TTC solution
Distilled water
Beaker
Bunsen burner
Tripod and gauze
4 Petri dishes with lids
Scalpel
White tile
Forceps
Clock
Wax pencil
Hand lens
Ruler

Abilities which could be assessed

Manipulative skills
Observation, identification, recording and interpretation

Investigation 10

Determine the approximate water potential of the cell sap of the beetroot tissue provided. Carry out the following instructions.

a Pipette out 7 cm^3 molar sucrose solution from the 20 cm^3 provided into a test tube and label as molar sucrose. From the remaining stock solution make up 7 cm^3 of M/2, M/3, M/4, and M/5 sucrose solutions and pour into the tubes, labelling each with its sucrose concentration. Make a note of your method of procedure and, if you have any doubts of its accuracy, consult the laboratory supervisor.

b Pipette 2 cm^3 of M sucrose into one test tube and label it, leaving 5 cm^3 M sucrose in the original tube. Place in a double test tube rack with the 2 cm^3 tube behind the 5 cm^3 tube. Repeat this procedure with the M/2, M/3, M/4 and M/5 solutions. You will now have two rows of five test tubes with all those containing 5 cm^3 in the front. Every tube should be labelled with its respective sucrose concentration.

c Cut the pieces of fresh beetroot provided into blocks approximately 1 cm square by 0.3 cm in depth. Leave these blocks on a tile and do not blot them. You will need 15 such blocks, all of approximately the same size, but exact measurement of each block is not necessary.

d You are now ready to start the experiment. *Note the time.* Place three of the beetroot blocks into each of the 2 cm^3 sucrose tubes, that is M, M/2, M/3, M/4 and M/5 concentrations. Manipulate the blocks so that they are just covered with the sucrose solution in each tube.

e Now wait for a period of at least 45 minutes.

f Draw a small amount of the red fluid from the 2 cm^3 M sucrose test tube containing the beetroot blocks. Now, *with great care,* introduce a single drop of this red fluid into the 5 cm^3 M sucrose solution. The drop should not go down the side of the tube but should be *released carefully in the centre of the liquid* about 5 mm below the surface. Note whether the red drop remains in the same place, or rises or sinks. Repeat with another drop and continue until you are quite certain you have made the correct observation about the behaviour of the drop.

g Repeat this procedure for the red fluid in the M/2, M/3, M/4 and M/5 solutions. In each case use a clean pipette.

The results

h Present your results in the form of a table.

i Explain very briefly why the red drop might rise or sink relative to the original solution.

j From your understanding of the factors that cause water to enter or to leave plant tissues, explain what you assume has been happening in each of the sucrose concentrations. In each case account for the observed behaviour of the red drops.

k What is the approximate water potential of the beetroot tissue in terms of sucrose molarity?

l Suggest ways in which improvements could be made in the experiment to give a more precise answer to *k* above. (OC)

Equipment per student

Pieces of fresh beetroot
20 cm^3 molar sucrose
50 cm^3 distilled water
5 cm^3 graduated pipette
12 test tubes and double-row test tube rack

Wax pencil
5 teat pipettes
Ruler
Clock
Scalpel
White tile

Abilities which could be assessed

Following instructions
Interpretation of data
Experimental design

Investigation 11

Determine the effects of the suspension K5 on the three sugars A, B and C. Sugar B is produced only in mammals.

K1 is a 0.2 molar solution of sugar A
K2 is a 0.2 molar solution of sugar B
K3 is a 0.2 molar solution of sugar C
K4 is a 0.4 molar solution of sugar C

Label 4 test tubes A, B, C1 and C2 respectively. Shake or stir the suspension K5 to mix it completely and put 10 cm^3 into each of the four tubes. Then:

to tube A add 10 cm^3 K1,
to tube B add 10 cm^3 K2,
to tube C1 add 10 cm^3 K3,
to tube C2 add 10 cm^3 K4.

Shake the tubes and place them in a container of water at 38–40°C. To each of the tubes fit the apparatus as shown in Fig 10.4 (only two of the tubes are shown). The tubes containing limewater can stand in a test tube rack in front of the water container.

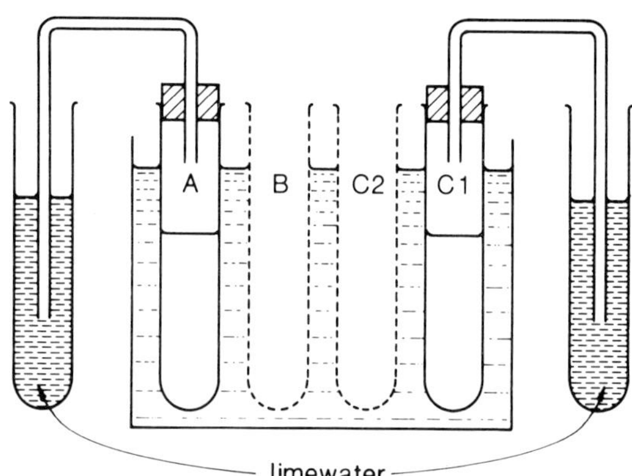

Fig 10.4 Assembled apparatus for Investigation 11

Now record the time. Leave this experiment until bubbles of gas are given off regularly (about 30–35 minutes). Maintain the temperature of the water at 38–40°C by adding more hot water when necessary.

a *i* When the bubbling is regular, count the number of bubbles given off in 1 minute from tube A. Now count the number of bubbles given off by tube C1 and then by tube C2 in 1 minute and record your results. Repeat this procedure to obtain 4 counts for each of the tubes A, C1 and C2. Record the average number of bubbles per minute for each of the tubes shown.

ii How many bubbles have been given off by tube B?

b Using only the reagents provided and the information gained from your experiment, identify as far as possible the sugars A, B and C (in K1, K2 and K3 respectively). Briefly record your methods, observations and conclusions.

c After you have counted the bubbles, carry out Benedict's or Fehling's test on the liquid in tube A. Record your observations and conclusion.

d What reaction has taken place in tube A to give the result you have observed in *c*? Give the name of this kind of reaction.

e Using similar apparatus, state briefly what you would do to support the hypothesis that enzymes are controlling the reactions in tube A.

f Give the name of the biological process taking place in tubes A, C1 and C2.

g Explain as fully as possible the similarities and differences between the rates of reaction in tubes A, C1 and C2.

h What would be the limiting factor in the reaction in tube C1?

i Explain or account for your observation on tube B.

j Why were the tubes allowed to stand for a time before taking readings?

k Place a very small drop of K5 on a slide. Add four drops of water and cover with a cover slip. Observe this under your microscope. Identify K5 with reasons, using all your observations.

l Briefly describe a simple procedure you could follow to determine whether K5 was living or dead.

(C)

Equipment per student

K5 50 cm^3 yeast suspension in water (4 g dried yeast in 100 cm^3 water)
K1 20 cm^3 0.2 M sucrose solution (ensure the absence of reducing sugar and keep refrigerated)
K2 20 cm^3 0.2 M lactose solution
K3 20 cm^3 0.2 M glucose solution
K4 20 cm^3 0.4 M glucose solution
Dilute hydrochloric acid
Solid sodium hydrogen carbonate
Benedict's solution
50 cm^3 limewater
Bunsen burner

12 test tubes and rack
Water bath
4 single-holed rubber bungs to fit test tubes each fitted with a delivery tube
10 cm^3 syringe
Stop clock
Microscope slides
Cover slips
Wax pencil
Thermometer
Microscope with light source

Abilities which could be assessed

Following instructions
Observation, identification, recording and interpretation
Interpretation of data
Experimental design

Investigation 12

Vessel X contains a 1% solution of soluble starch. Vessel Y contains an enzyme preparation. Pour 8 cm^3 of starch solution into a boiling tube and 4 cm^3 of the enzyme suspension into another tube. Place the tubes in a water bath at about 35°C for 5 minutes to equilibrate. Then mix the contents of the two tubes. At 2 minute intervals remove a few drops of the mixture with a drop pipette and add to a small quantity of iodine solution on the white spotting tile. Record carefully the colour of the mixture at each time interval. Devise a suitable control experiment. Repeat the above procedure to determine the effect of the sap from the germinating barley grains on solution X.

Grind thoroughly about 25 of the grains provided in about 10 cm^3 of distilled water in a mortar. Filter off the liquid through two layers of muslin. Centrifuge to sediment the starch.

In both experiments record your observations and offer an explanation for them.

Why and in what ways are the two experiments similar?

Credit will be given for details of experimental technique, relevant qualitative tests and presentation of results.

(O)

Equipment per student

25 germinating barley grains — sow over succeeding days on damp blotting paper for 2–5 days before the practical
50 cm^3 1% Analar starch solution (X)
20 cm^3 0.5% Analar amylase solution (Y)
Iodine in potassium iodide solution
Distilled water
Pestle and mortar
Access to a centrifuge
6 boiling tubes and rack
White tile
Stop clock
Water bath
2 graduated 10 cm^3 pipettes
Teat pipettes
Filter funnel and muslin
Beaker
Thermometer

Abilities which could be assessed

Following instructions
Interpretation of data
Experimental design

Investigation 13

a Half fill a test tube with milk and add dilute hydrochloric acid drop by drop, shaking after each addition, until precipitation is complete. Filter, retaining the filtrate, and remove excess water from the precipitate by pressing it between filter papers. Put some of this precipitate in a test tube and add about 2 cm^3 of alcohol. Shake gently, allow to stand for a few moments, then pour off the alcoholic solution into a test tube half full of cold water. Note and record what happens, and state what conclusions you draw.

Carry out tests for both protein and reducing sugar on both the precipitate and the filtrate to find out as much as you can about their composition. Record your results and conclusions.

b Half fill another tube with milk, and put it in a water bath at 40°C. When the milk has reached this temperature, add 5 drops of the enzyme solution A and mix well. Put the tube back in the bath and keep its temperature as near as possible to 40°C, examining the contents of the tube at intervals. After 10 minutes, take it out and examine its contents. Then shake it vigorously and allow to stand. Record the appearance before and after shaking.

Bearing in mind the deductions you made from the tests carried out in *a*, state what effects you think the acid and the enzyme each had on the milk, pointing out how these effects are similar and how they differ. (You may, if you wish, carry out tests on the precipitate obtained in the second section.)

(JMB)

Equipment per student

A Rennet in a dropping bottle
Fresh milk
Dilute hydrochloric acid
98% alcohol
Biuret test reagents
Benedict's solution
Test tubes and rack
Tripod and gauze
Bunsen burner
Beakers
Filter funnel
Filter papers
Thermometer

Abilities which could be assessed

Following instructions
Interpretation of data

Investigation 14

L1, L2, L3 and L4 are four liquids of which one is a 1% solution of glucose, one a 1% solution of sucrose, one a 1% starch solution and one is water. Use the enzyme solution (invertase) provided and any other necessary reagents to identify the four liquids.

Write a brief explanatory note of your method and results.

(W)

Equipment per student

L1 1% sucrose solution
L2 1% starch solution
L3 Distilled water
L4 1% glucose solution
Invertase solution
Benedict's solution
Dilute hydrochloric acid
Solid sodium hydrogen carbonate
Iodine in potassium iodide solution
Test tubes and rack
Bunsen burner
Wax pencil
Teat pipettes
Water bath at 35–40°C

Abilities which could be assessed

Manipulative skills
Experimental design

Appendix 1

Lists of abilities to be assessed as specified by the GCE examining boards requiring teacher assessment of practical work in Advanced level biology

The Associated Examining Board

Abilities to be assessed
Four areas of practical ability will be assessed and all four will be weighted equally and marked on the six point scale of 0–5, (to be replaced by 10 areas of practical ability in 1985).

A The ability to apply experimental techniques
Skills to be assessed under this heading might include:
Setting up apparatus independently, following a list of instructions; equipment might include light microscope set up for optimum use, spirometer, colorimeter, pH meter, kymograph, centrifuge
Applying suitable techniques with relevant manipulative skills, such as chromatography, simple slide preparations with appropriate staining, titration, weighing and measuring
Separating by dissection and displaying morphological, histological and anatomical features of plants and animals

Note: These techniques and skills are not to be used as an end in themselves, but should form part of a valid biological investigation.

B The ability to observe and record results of investigations
Skills to be assessed under this heading might include:
Recording observations accurately and correctly
Reading measuring instruments accurately
Drawing with adequate precision, showing relevant information
Recording unexpected or unusual aspects of an investigation

C The ability to interpret observations and draw conclusions
Skills to be assessed under this heading might include:
Recognising and allowing for errors of measurement, the source of experimental errors
Interpreting results, bearing in mind their accuracy and likely statistical significance
Evaluating the findings correctly
Drawing the correct conclusions from results
Showing positive scepticism of results and pointing out drawbacks of methodology
Suggesting further areas of relevant research

D The ability to plan and carry out investigations
Skills to be assessed under this heading might include:
Recognising the problems and formulating valid, testable hypotheses
Choosing appropriate experimental methods and sampling techniques
Choosing appropriate apparatus
Recognising the need for control(s), recognising sources of error and effects of these on outcomes
Modifying experimental method after initial work or unexpected outcomes
Producing a work plan that is logical and sequential

Joint Matriculation Board

Abilities to be assessed
Each of the four abilitites (A, B, C and D) is to be assessed on a ten-point scale (1 to 10).

A Possession of appropriate manipulative skills
There are four skills to be assessed under this heading. It is not intended that these skills should necessarily be assessed as separate exercises.
1 The use of dissection instruments
2 Cutting sections, mounting and staining of temporary preparations

3 The use of a lens and a microscope (both low and high magnification)
4 The handling of apparatus

B Carrying out observational investigations
Under this heading the student is to be assessed on the ability to exercise powers of observation, to identify, to interpret and record important features both microscopically and macroscopically and, where appropriate, to make comparisons. The skills listed under A will be involved in the investigations, and care must be taken to avoid confusing assessments of the two categories.

C Carrying out experimental investigations
Under this heading the student is to be assessed on:
1 The ability to carry out investigations in accordance with specified procedures
2 The presentation and handling of results
3 The interpretation of results

In handling data, the student should be required to draw quantitative conclusions from the data by making simple calculations. In handling data, it is preferable that the student should use results obtained within the class as often as possible. In interpreting class results, the student must keep in mind the actual situations and constraints that this imposes on what conclusions might be drawn.

D Planning investigations
The assessment should be based on the student's ability to plan an investigation with proper regard for the limitations of the methods proposed, the need for controls, the choice of a form of presentation for the results, and the weight to be attached to such results.

This ability is not intended to be interpreted as the extent to which a student can satisfactorily design and carry out an independent research project.

Joint Matriculation Board in association with the other GCE Examining Boards Biology (Nuffield) (Advanced)

Abilities to be assessed
The assessment of students' practical abilities should be made under three Operational Divisions and on a ten point scale (1–10).

Procedure
All those activities displayed by the student when in action at the laboratory bench or in the field. These include the care and competence shown in
1 selecting a suitable approach
2 carrying out operations
3 making observations and measurements

A reasonable overall competence should be sought rather than the development of special manual skills.

Recording
Methods used for recording observations and experimental results should be taken into account, eg short notes, sketches, diagrams, tabulation of results, graphs. The extent to which the students' records correspond to their actual results should also be considered. The student should have indicated possible errors resulting from the materials and methods used.

Handling of results
Ability to
1 analyse and interpret data
2 draw conclusions
3 formulate further hypotheses where appropriate

Appendix 2

Specification, in the form of a hierarchic taxonomy, of the objectives of practical work with suggestions of methods by which assessment could be undertaken

A Knowledge of apparatus
All that would be required under this category would be the ability to name practical apparatus and to state its purpose in terms of use. Assessment could be undertaken by presenting the pupil with pieces of apparatus and requiring him or her to name them and to describe their use.

B Knowledge of procedures
All that would be required here is basic knowledge of routine procedures, ie the extent to which the pupil knows the procedures for carrying out routine practical operations which are basic to the subject. Assessment could be undertaken by using questions requiring descriptions of the procedures.

C Knowledge of ways of using apparatus
Under this heading the pupil would be expected to know how to use apparatus involved in carrying out routine procedures and how to handle the apparatus to achieve varying degrees of accuracy. Knowledge and use of safety precautions would also be included.

D The ability to use apparatus
Under this heading the pupil would be required to show that he could use apparatus of which he had knowledge. This means that he should be able to combine the aspects of knowledge required under A, B and C and to apply some degree of manipulative dexterity in the performance of relevant operations.

Assessment for these last two categories could be undertaken by means of simple exercises involving the use of the particular piece of apparatus. The criteria for judging the quality could be based either on pre-determined results with an allowed margin of error and/or direct observation of the work being carried out.

E The ability to implement procedures
This heading would involve the carrying out of those procedures of which the pupil had knowledge (***B***).

Assessment could be undertaken by providing simple instructions for the particular procedure and requiring the pupil to perform it. The assessment would take into account the extent to which the pupil carried out any checks necessary to ensure the satisfactory working of the apparatus used and the necessary accuracy of results.

F The ability to select appropriate procedures for a particular practical problem
Under all the previous headings the apparatus or procedures have been in the limited context of a specific or specified purpose. This heading involves the ability of the pupil to select, by applying the appropriate criteria, the most suitable apparatus and/or procedure for a particular practical or experimental task.

Assessment could be undertaken by presenting the pupil with a practical problem and a range of alternative apparatus and procedures. The extent to which the most appropriate choices are made would form the basis of the assessment.

G The ability to observe the material under investigation
Under this one heading there are two levels of complexity. At its simplest level the ability is concerned with the identification and classification into known categories of the objects or processes which are being investigated, ie a descriptive skill based on observation. At a more complex level the ability involves the pupil in the exercise of a degree of discernment in establishing patterns from his observations and systematising the information derived from them.

Assessment could be undertaken by requiring the pupil to translate his observations into oral or written terms. The criteria adopted for the measurement of the ability would be the extent to which the pupil's observations were comprehensive, systematic and ordered in terms of significance.

H The ability to observe changes or differences taking place in the material under investigation
This heading is an extension of ***G***. There, attention was to be paid to the general method and effectiveness of the pupil's observation; under this heading is included the extent to which the pupil is able to recognise the changes that take place in the material being studied and, having recognised such changes, to take whatever steps are necessary to examine them systematically.

Assessment could be undertaken by presenting the pupil with a practical situation and requiring him to observe the changes which take place and to identify and isolate the changing factors in such a way that they can be studied more comprehensively, eg by refined counting techniques and sampling procedures and by eliminating possible errors and variations resulting from the techniciques used or from inherent variation in the material being studied.

I The ability to record appropriately observed material and the changes which take place in it
This heading covers the ability of the pupil to make and keep records of the activities described in ***G*** and ***H***.

Assessment could be undertaken by evaluating the use made by the pupil of the available methods of recording observations. The criteria upon which the assessment is made should include not only the pupil's selection of the available methods of recording but also the extent to which the pupil is aware of possible distortions of the data by the use of different methods of presentation. They should also include the extent to which conclusions drawn from the data arise from the data itself or are conditioned by the method of presentation.

J The ability to devise new apparatus or techniques to meet the demands of a particular problem
This heading involves the ability of the pupil, when faced with a practical problem which cannot be solved satisfactorily by the use of familiar apparatus and procedures, to make modifications and adaptations to known apparatus and procedures to meet the demands of the new problem.

Assessment here requires that the pupil be presented with a situation in which the apparatus and techniques used and the problem posed are sufficiently within his experience for him to appreciate that the former are insufficient to solve the latter.

K The ability to plan and carry out a practical investigation
This heading involves all the preceding categories and the use of all the pupil's practical experience and skill in the design of practical work and its execution.

Assessment can be undertaken by presenting the pupil with a problem and requiring him to plan an appropriate practical procedure to solve it. The initial plan should be written out by the pupil and evaluated before it is implemented. The assessment should take into consideration the extent to which the pupil has anticipated all the problems which he could justifiably be expected to anticipate at the planning stage. It should also include the extent to which the pupil recognises problems as they arise and modifies his plan to overcome them. It is therefore not necessary to ensure that the plan is foolproof before the exercise begins, but care must be taken to ensure that a plan which has serious faults when first designed is modified to enable the pupil to proceed with the practical work.

L Attitudes to practical work
In almost all subjects in which it is felt necessary to incorporate practical work in the overall pattern of assessment, one of the most important educational objectives is the establishment of particular attitudes towards practical work. The desirable attitudes here are, for example, 'willingness to co-operate in the normal routine of a laboratory', 'persistence', 'resourcefulness', 'enthusiasm', 'the ability to work as a member of a group', 'commitment to practical work as a worthwhile pursuit without compulsion'.

Attempts to assess such qualities are rarely included in public examinations because of the very real problems of making assessments with any degree of objectivity. These qualities, moreover, are those upon which it is particularly difficult for individuals to reach agreement over their recognition and definition and upon the standards to be applied in assessment if they can be identified. It is, however, necessary to note that this area of attitudes is probably the most important of all in terms of teaching objectives. Careful consideration should, therefore, be given to the matter before a decision is taken to include or exclude what in Bloom's terminology are 'affective' objectives.

Appendix 3

Books used by teachers as sources of assessed practicals and data exercises

ASE Laboratory Books (1973) *Plant Physiology; Cytology, Genetics and Evolution.* John Murray.
Baron, W.M.M. (3rd Ed. 1979) *Organisation in Plants.* Edward Arnold.
Brocklehurst, K. & Fielden, P.S. (1974) *Biology for Modern Courses.* English Universities Press.
Brown, G.D. & Creedy, J. (1970) *Experimental Biology Manual.* Heinemann.
Clarke, R.A. et al (1968) *Biology by Inquiry Books 1, 2 and 3.* Heinemann.
Clarke, W.M. & Richards, M.M. (1970) *The Locust as a Typical Insect.* John Murray.
Harrison, D (1970) *Problems in Genetics.* Addison-Wesley.
Head, J.J. (1972) *Discovering Biology.* Oxford University Press.
Kirby, T.W. & Clarke, H.P. (1976) *Experimental Biology — A Practical Handbook.* Oxford University Press.
Lewis, T. & Taylor, L.R. (1967) *Introduction to Experimental Ecology.* Academic Press.
Luker, A.J. & Luker, H.S. (1971) *Laboratory Exercises in Zoology.* Butterworth.
Mackean, D.G. (1971-83) *Experimental Work in Biology.* John Murray. (nine titles)
Mackean, D.G., Worsley, C.J. & Worsley, P.C.G. (1983) *Class Experiments in Biology.* John Murray.
Nuffield Advanced Biological Science (1970) *Laboratory Guides. Teachers' Guides I and II. Study Guide and Teachers' Guide III.* Penguin Education.
Open University *Biology Course Units.* Open University Press.
Revised Nuffield Biology (1975) *Texts 2, 3 and 4. Teachers' Guides 2, 3 and 4.* Longman.
Roberts, M.B.V. (1974) *Biology, a Functional Approach: Students' Manual.* Nelson.
Sands, M.K. (1971) *Problems in Plant Physiology.* John Murray.
Sands, M.K. (1975) *Problems in Animal Physiology.* John Murray.
Sands, M.K. (1978) *Problems in Ecology.* Bell & Hyman.
Schools Council (1977) *Micro-organisms.* Hodder & Stoughton.
School Science Review, John Murray.
Torrance, J. (1981) *Interpretation Tests in Biology.* Edward Arnold.
Tranter, J. (1978) *A-level Biology Structured Questions.* Blackie.
Williams, J.I. & Shaw, M. (1976) *Micro-organisms.* Bell & Hyman.
Wratten, S.D. & Fry, G.L.A. (1980) *Field and Laboratory Exercises in Ecology.* Edward Arnold.

References and further reading

DES/APU (1981) *Science in Schools. Age 11: Report No 1*. London: HMSO.
DES/APU (1982) *Science in Schools. Age 13: Report No 1*. London: HMSO.
DES/APU (1982) *Science in Schools. Age 15: Report No 1*. London: HMSO.
Eggleston, J.F. & Kerr, J.F. (1969) *Studies in Assessment*. London: English Universities Press.
Kelly, P.J. & Lister, R.E. (1969) Assessing Practical Ability in Nuffield A-level Biology. In Eggleston, J.F. & Kerr, J.F. (eds) *Studies in Assessment*. London: English Universities Press.
Kerr, J.F. (1963) *Practical Work in School Science*. Leicester University Press.
Macintosh, H.G. (ed.) (1974) *Techniques and Problems of Assessment*. London: Edward Arnold.
Sands, M.K. (1981) *Teacher Assessment of Practical Work in JMB A-level Biology*. Manchester: Joint Matriculation Board.
Schools Council (1965) *Examinations Bulletin No 5. School Based Examinations. Examining, Assessing and Moderating by Teachers*. London: HMSO.
Thompson, J.J. (ed.) (1975) *Practical Work in Sixth-form Science*. University of Oxford.
Whittaker, R.J. (1974) The Assessment of Practical Work. In Macintosh, H.G. (ed.) *Techniques and Problems of Assessment*. London: Edward Arnold.